SpringerBriefs in Materials

For further volumes:
http://www.springer.com/series/10111

Nicholas Travis Kirkland
Nick Birbilis

Magnesium Biomaterials

Design, Testing, and Best Practice

 Springer

Nicholas Travis Kirkland
Department of Advanced Technology and
    Science for Sustainable Development
Nagasaki University
Nagasaki
Japan

Nick Birbilis
Department of Materials Engineering
Monash University
Clayton
Australia

Additional material to this book can be downloaded from http://extras.springer.com/

ISSN 2192-1091   ISSN 2192-1105 (electronic)
ISBN 978-3-319-02122-5   ISBN 978-3-319-02123-2 (eBook)
DOI 10.1007/978-3-319-02123-2
Springer Cham Heidelberg New York Dordrecht London

Library of Congress Control Number: 2013949245

Printed on acid-free paper

Springer is part of Springer Science+Business Media (www.springer.com)

Foreword

Orthopaedic surgery today depends heavily on the development and use of bio-materials. The best known examples are implants used for fixation of fractures and those used in joint replacement. In these settings biomaterials contribute greatly to the improvement of the health and wellbeing of humankind. Biomaterials for orthopaedic surgery are an important and growing area in this rapidly changing clinical discipline. New and improved materials are required to enhance device performance, to improve function, deliver bioactive compounds and achieve the goal of tissue regeneration.

In order for orthopaedic implants to achieve their desired functions in humans, a number of key factors must be addressed. From the mechanical standpoint an implant material must have the properties to suit the desired function. The bio-material will necessarily interface and interact with the host cells and tissues and therefore must be compatible with the environment of the living human. Due to the rapid development of putative biomaterials, there is a pressing need for compre-hensive research with respect to materials properties, biocompatibility, toxicity, fitness for purpose and stability over time in the intended environment.

The area of biodegradable magnesium implants is one such emerging tech-nology. There are many aspects of biodegradable magnesium materials and implants which are in need of further research. One such area, prior to further trialling and eventual clinical implementation, is the best-practice execution of in vitro tests. This monograph addresses this issue, and may serve as a platform for understanding the aspects which relate the material/alloy with its degradation.

Canberra, June 2013

Prof. Paul Smith
Director of the Department
of Clinical Orthopaedic Surgery

Preface

This book presents a snapshot of a rapidly growing area of research, where although significant progress has recently been made, many substantial challenges remain. The number of articles being published in the field of magnesium biomaterials is continually rising, as seen by the Scopus citation information (www.scopus.com) presented below. As more researchers join the field, a consolidated monograph which summarises the key issues and identifies the main challenges is timely. This book is also a first step towards standardising tests and making researchers more aware of the subtleties involved in testing magnesium biomaterials. Our hope is that future studies will allow comparisons between studies to be more meaningful. The field of magnesium biomaterials is indeed interdisciplinary, embracing a diverse range of more mature fields, each of which requires their own specific skill set. For example, alloy designers are often unaware of organic components in media, and biologists can struggle to understand electrochemistry. This monograph aims to provide some guidance and the necessary unified platform to be able to conduct practical, clinically relevant, in vitro testing of biodegradable magnesium biomaterials. The book is intended for investigators in the area of metallic implants, for existing researchers wishing to transition to this evolving field, and for graduate students who conduct practical experiments in the laboratory.

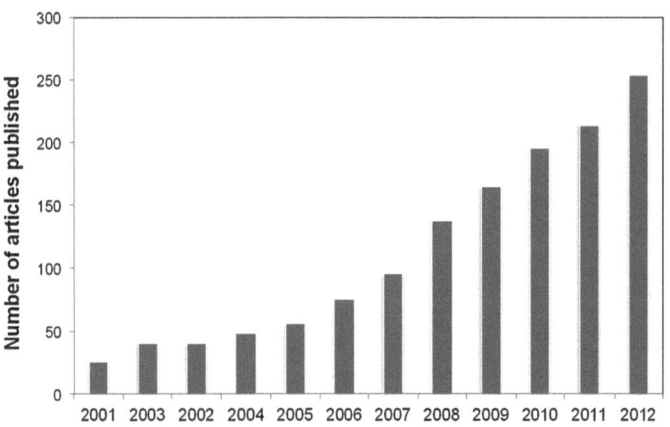

This book would not have been possible without the support of our institutions, to which we are grateful. The compilation of the manuscript in Nagasaki will always bring fond memories. Thanks for assistance with figures and discussion to the following: Jake "Diablo" Cao, Kateryna Gusieva, Jay Waterman and Thanh Loan Nguyen. A special thanks is extended to Kate Nairn for her assistance in proofreading. Finally, a big thanks for the support of our friends and family.

Nicholas Travis Kirkland
Nick Birbilis

Contents

Abbreviations

AA	Amino acids
BMG	Bulk metallic glass
BSA	Bovine serum albumin
BSS	Balanced salt solutions
CPE	Constant phase element
E_{corr}	Corrosion potential
EBSD	Electron backscattered diffraction
EBSS	Earle's balanced salt solution
EDL	Electrical double layer
EDS	Energy-dispersive X-ray spectroscopy
EIS	Electrochemical impedance spectroscopy
FTIR	Fourier transform infrared spectroscopy
G	Gibbs free energy
H_2^{evo}	Hydrogen evolution experiment
HBSS	Hanks' balanced salt solution
HP	Human plasma
i_{corr}	Corrosion current density
KBM	Kirkland's biocorrosion medium
MEM	Minimum essential medium
MEM_{BSA}	MEM with bovine serum albumin
MEM_{FBS}	MEM with foetal bovine serum
ML	Mass loss experiment
OCP	Open circuit potential
PBS	Phosphate buffered saline
PDP	Potentiodynamic polarisation
R_{CT}	Charge transfer resistance
R_f	Film resistance
R_{tot}	Total resistance
R_{P-B}	Pilling-Bedworth ratio
SB_{CO2}	Sodium bicarbonate with CO_2 environment buffer
SEM	Scanning electron microscope
T_{phy}	Physiological temperature
T_{rm}	Room temperature

Chapter 1
Introduction to Magnesium Biomaterials

Abstract Over the past two decades, research focused on magnesium (Mg) and its alloys as potential biomaterials has progressed rapidly. This is not surprising, given the unique advantages of Mg alloys as biomaterials, especially their combination of good mechanical properties and safe biodegradation. Current investigations examine a wide range of properties, from mechanical strength to toxicity; however, perhaps the most elusive aspect thus far is the biocorrosion of Mg alloys. This chapter provides a succinct review of the area, highlighting salient features and aspects of Mg biomaterials and their development to date.

Keywords Magnesium · Biodegradable · Implant · Biomaterial · Metal · Bone replacement · Orthopaedic · Alloys · In vitro · Biocorrosion

1.1 Changing the Way We Look at Biomaterials

1.1.1 Background to Biomaterials

Biomaterials have been used in various forms in societies around the world for thousands of years. The Aztecs, Romans and ancient Chinese all used gold in dentistry, while glass eyes have appeared in historical references for almost as long [1]. However, choice of materials and designs remained virtually unchanged until the turn of the last century. Enhanced scientific knowledge and advanced production have resulted in a number of new and improved biomaterials, and their applications are increasing dramatically. An example of this changing approach to biomaterials is the use of parachute cloth as a vascular prosthesis, carried out after World War II [2]. It was also in the twentieth century that orthopaedic biomaterials truly started to be clinically tested and understood. Experiments such as John Charnley's use of polytetrafluoroethylene (PTFE) and stainless steel for total hip replacement helped improve the human condition and encouraged further interest in biomaterials for

N. T. Kirkland and N. Birbilis, *Magnesium Biomaterials*, SpringerBriefs in Materials, DOI: 10.1007/978-3-319-02123-2_1, © The Author(s) 2014

orthopaedic replacements [3, 4]. Developments in this area will continue to increase, as biomedical departments and industrial interest are established around the world, and our understanding of biological interactions improves.

According to a survey by Markets and Markets, the global biomaterials market is set to reach almost US \$65 billion by 2015, up from US \$25.6 billion in 2008 [5]. With an annual growth of over 15 % year-on-year, it is one of the fastest growing economic sectors. As people live longer and wish to remain active, biomaterials technology must keep pace.

1.1.2 Traditional Implants: Strengths and Weaknesses

There are a number of different biomaterials currently available; these can be broadly classified into four groups, shown in Table 1.1.

The full discussion of each of these materials would require far more space than this monograph permits and would be out of scope; however, it suffices to say that each of these materials has unique benefits and challenges, and thus each is used for the applications that best suit their individual properties.

1.1.3 The Move Towards BioDesigned Materials

The first biomaterials were chosen primarily because they were bioinert (i.e. they did not seem to result in any positive nor negative response from the host). Although acceptable for short periods of time, these materials had several drawbacks, including relatively slow healing times and a higher rejection rate over long periods [6]. Following this, biocompatible or biodegradable materials were developed, which could speed the rate of the healing and only provide functionality for a required time, after which the material would safely degrade in the body. These materials include degradable polymers such as PMMA derivatives and minerals including those based on hydroxyapatite. Recently, a third generation of biomaterials has been described by Hench and Polack [7] who proposed that by combining the features of biocompatibility and biodegradability to enhance the growth of specific cells while providing the necessary functions for only as long as needed, more functional biomaterials can ensue.

The demand for short-term, non-permanent implants is set to increase as the general population ages. The advancement of implants from bioinert to biocompatible/bioabsorbable to those designed to stimulate responses/promote growth has thus far been mostly limited to polymers. However, with the right design and coatings, Mg-based biomaterials can be part of this new generation of functional, resorbable biomaterials—particularly in orthopaedic applications.

Table 1.1 Strengths and weaknesses of current biomaterials

Material	Applications	Advantages	Disadvantages

Polymers
Dacron
	• Blood vessels
	• Ligament
	  replacement
		• Ease of fabrication
		• Resilience
			• Low compressive
			  Strength/hardness

Nylon
Teflon
PMMA
Polyethylene
	• Ear & nose
	  replacement
	• Soft tissue
	  applications
	• Bone cements
	• Drug delivery
	• Bone regeneration
		• Processable
		• Low cost
		• Partly biodegradable
			• Deformation over time
			•Degradation

Ceramics
Alumina
Hydroxyapatite
Carbides
PSZ
	• Dental implants
	• Hip sockets
	•Maxillofacial
	  reconstruction
		• High compressive
		  strength
		• Biocompatible
		• Bone conducting
			• Difficult to manufacture
			• Brittle
			• Not resilient

Composites Carbon–carbon	• Heart valves • Joint implants	• Mechanical strength	• Difficult to manufacture

Metals
Stainless steel
  (316/L)
HA-coated Ti
Co-Cr-Mo alloy
Gold
	• Joint replacement
	• Screws
	• Hip nails
	• Bone fixation
	• Maxillofacial
	  applications
		• Mechanical strength
		• Wear resistance
		• Ductility
			• Corrodes (metal ions)
			• Dense
			• High modulus (stress
			  shielding)

Data adapted from [6, 33, 34]

1.2 Mg Alloys: Bridging the Weaknesses

1.2.1 History of Magnesium

By molarity, Mg is the eighth most abundant element in the world, constituting approximately 2.5 % of the earth's crust [8]. Although first discovered by Joseph Black in 1755, it was first fully isolated by Sir Humphrey Davy in 1808, who created Mg metal electrolytically. In 1931, Antoine Bussy prepared Mg by heating magnesium chloride (MgCl) with potassium (K), to precipitate Mg and KCl [9]. Several years later, Michael Faraday reduced dehydrated MgCl by electrolysis, obtaining pure Mg [10]. Mg production is now a >1.17 Mton/yr industry [11].

Historically, Mg usage has been limited by three factors: (1) difficulties with isolation, (2) rapid corrosion and (3) relatively poor mechanical properties (including anisotropy, low creep strength and low ductility). As such, Mg was initially ignored by industry and considered primarily a curiosity [12]. It remained in production principally for use as the main alloying element in the majority of aluminium alloys (in which it improves a range of properties). However, the advent of aviation and World Wars resulted in an increase in interest in Mg, where its low density (1.738 g/cm^2) and moderate strength enabled it to be employed in several aircraft.

1.2.2 Potential of Magnesium as a Biomaterial

Reports of Mg as a potential biodegradable implant have existed for more than a century [13]. Early use as a ligature, plate and as screws displayed varying levels of success, although most failed due to low purity levels or poor understanding of corrosion behaviour [13–19]. The rapid corrosion that took place and associated release of hydrogen gas meant that few studies were performed in the middle of the last century, as many surgeons put Mg to one side. However, recent increases in Mg purity and better understanding of corrosion kinetics have resulted in somewhat of a boom in research, starting in the early 2000s, with many groups around the world now studying modern Mg alloys for bioapplications. The drastic increase in research is due to the potential of Mg, particularly in orthopaedic applications. The fourth most abundant metal ion found within the human body, Mg^{2+}, is a vital nutrient for life and is present in every cell type for all organisms [20–25]. Mg is a fundamental dietary requirement and is efficiently controlled in the body by homeostatic mechanisms [26, 27]. As a result, toxicity of Mg is generally not considered to be a problem.

As outlined in Table 1.2, Mg and its alloys offer a number of other benefits over many current implant materials, including low density, high damping capacity, ease of machinability, elastic modulus close to that of bone, biocompatibility and osteogenicity. However, what truly separates Mg from current metallic implants is its ability to biodegrade in vivo. This creates the potential for an implant to last as long as its intended function, without the need for surgical retrieval. In orthopaedic applications, existing implants are dominated by load-bearing designs based on permanent metals, such as stainless steel and titanium. With Mg, the benefits of a temporary, biodegradable approach to implant design include saving the patient and healthcare provider (it saves it for both the patient and the provider) money, time and any potential complications associated with implant removal or revision.

1.3 Issues Facing Mg Biomaterial Development

1.3.1 Primary Drawbacks of Mg Biomaterials

All implant materials are exposed to a relatively harsh environment in the human body—an oxygenated solution with a salt (principally NaCl) content of ~ 0.9 wt. % at a pH of ~ 7.4 and a temperature of ~ 37.1 °C [28]. Bodily fluids which contain water, dissolved oxygen, complex compounds, sodium, potassium, calcium, magnesium, phosphate and sulphate ions, amino acids, proteins, plasma and a number of other substances make interactions and corrosion processes complicated [28]. Moreover, the in vivo environment can change dramatically in the immediate area of an implant after surgery; these changes also need to be considered when designing and testing biomaterials. The unique environment and associated issues

Table 1.2 Key benefits of Mg biomaterials

Benefit	Details
Low density/ High specific strength	Mg density (1.738 g/cm^3) [35] is close to that of cortical bone (1.75–2.1 g/cm^3) [36]. Pure Mg has a strength-to-weight ratio of approximately 130 kN•m/kg
High damping capacity	Mg has the highest damping capacity (ability to absorb energy), of any metal [37]. This can be important in load-bearing applications, where the shock- and vibration-absorbing properties of Mg could provide significant benefit over other materials
Machinability and dimensional stability	Mg is the easiest structural metal to machine, and stable final dimensions are easy to achieve [38]. Consequently, complex shapes are easily producible, which is crucial for the often intricate shapes that are required for medical applications [39–41]
Reduction in likelihood of stress shielding	Stress shielding is the process by which bone mass and density will decrease in the vicinity of an implant with a mismatched (usually higher) stiffness value, as it transfers the load away from the adjacent bone. This can cause serious problems and implant failure if it continues [42–45] and is known to be a problem with current orthopaedic devices based on stainless steel or titanium, which have a density, elastic modulus and yield strength in an order of magnitude higher than that of bone [46]. Pure Mg has an elastic modulus of ~45 GPa, which is much closer to that of human cortical bone (~20 GPa) than most common Ti alloys (110–120 GPa) [20]. Combined with a density very close to that of bone, stress-shielding-related problems can be greatly reduced for many orthopaedic implants, especially those in high-load-bearing areas
Biocompatibility and osteogenesis	Current biomaterials such as pure Ti are relatively inert in the body, meaning they exhibit little host response, positive or negative [47, 48]. In contrast, Mg is considered biocompatible [21, 26, 49–52] and has been shown to increase the rate of bone formation [50, 53]. Mg is known to have a positive influence on bone fragility and strength [50, 54, 55]
Safe Degradation	Titanium, stainless steel and Co-Cr implants are not designed to degrade safely in the body. However, all surgically implanted metal alloys undergo some electrochemical degradation in the complex and corrosive environment of the body [28]. Combined with significant wear that can occur in load-bearing applications, particles of the implant can be released into the surrounding tissues, causing discomfort and potential health risks [56, 57]. Although the bulk material may be considered bio-inert, the way in which the particles are metabolised within the body can lead to acute inflammation and eventually implant failure [58]. Mg minimises all these issues. The gradual release of Mg ions in the body is dealt with effectively [21, 26, 49–52]. The corrosion of Mg in the body would eventually result in complete degradation. This also means that patients would benefit from only temporary exposure to a "foreign" object in their body. This avoids the complications that can and do occur over time for many implants; issues are more likely to arise the longer an implant remains in vivo [59]

that Mg biomaterials must overcome are outlined in Table 1.3. These issues may all be mitigated, however, through the use of alloying, coating and design (as discussed in Chap. 4). A discussion on the benefits and drawbacks of Mg biomaterials may be found in [29].

1.3.2 The Complex Nature of Biocorrosion

Of the issues facing the use of Mg alloys as a biomaterial, perhaps the greatest arises as a result of corrosion itself. The degradation of the alloy has second-order effects. If the material degrades too quickly, toxicity, mechanical degradation and hydrogen evolution become problems. The management and mitigation of these issues require an interdisciplinary approach, involving metallurgists, toxicologists, corrosion scientists, modellers and surgeons. The need for a large team with widespread knowledge of many areas has perhaps been the largest barrier to the rapid development of Mg-based biomaterials.

1.3.3 The Need for In Vitro Testing

New biomaterials are initially investigated using synthetic systems that, to some extent, mimic the environment within the body. Although animal experiments will certainly be necessary before a final product is produced, it is both unrealistic and unethical to expect a large number of studies to be carried out in vivo if such basic aspects as toxicity of alloying elements and reliability in long-term in vitro tests have not been fully addressed. Consequently, the first line of experiments for any potential Mg-based biomaterial will be those based outside a living organism. To date, there are no standards available that are suitable for the study of Mg biomaterial corrosion. Existing standards used for corrosion of stainless steels and other metals have been found to be inappropriate for the relatively fast degradation of Mg [30]. In addition, they do not take into account the complicated simulated body fluids (SBFs) that are required to properly test Mg for bioapplications. This lack of benchmarks means that it is critical to understand both the experiments used to study Mg biocorrosion, and the effects of the various parameters on the results of those experiments.

1.4 Future Development of Mg Biomaterials

Even if a Mg alloy is found to degrade completely safely in the body, the implant will be changing shape and mechanical properties over its entire life, adding a significant complication to a complete life-cycle design. A significant step in

Table 1.3 Potential drawbacks of Mg biomaterials

Drawback	Details
Low elastic modulus	Although the lower elastic modulus of Mg may be beneficial with respect to stress shielding, it also means that there may be a greater chance of failure in high-load applications, such as in the spine, where compressive loads during certain activities may exceed 3,500 N [60]. It is vital to ensure that any implant is designed to sustain its load without deformation. However, this aspect is even more crucial when considering degradable materials, as appropriate mechanical support is required throughout the entire bioresorption and bone remodelling process
Rapid degradation	Mg implants are intended to completely degrade, but at a slow rate that reduces H_2 gas formation and is similar to the pace of bone remodelling. The rapid degradation of Mg implants observed early in the last century [14, 17] has been greatly reduced by recent advances in controlling the purity of Mg and alloying elements; however, rapid degradation can still be an issue [61–63]
Resorption problems	The rapid degradation of Mg alloys may cause an adverse biological response as Mg and other element ions are released too quickly into the surrounding tissues. All of the alloying elements will eventually enter the patient and must be selected with non-toxicity as a primary factor. However, elements normally present in the body (e.g. Zn, Ca, Mn) can also be toxic if the release rate is too high, as the levels cannot be dealt with appropriately (e.g. excess Mg via kidneys or hydrogen gas via soft tissues). Thus, a truly biocompatible Mg alloy must both avoid the use of toxic alloying elements and ensure appropriate release rates for other elements, even those elements which naturally occur in the body
Hydrogen evolution	The release of H_2 and subsequent cyst formation following implantation of Mg can cause various problems. Gas pockets may form next to the implant and cause separation of tissue and/or tissue layers [22, 64]. H_2 bubbles may delay healing at the surgical site, leading to necrosis of surrounding tissue [65]. In the worst case scenario, gas bubbles could block the blood stream, causing death [66]. If the degradation is too rapid, the amount of H_2 produced will accumulate because it cannot diffuse through the surrounding soft tissues at a sufficient rate [67]
	The in vitro hydrogen evolution rates for various Mg alloys containing Zn, Al and Mn are reported to be within tolerable rates (i.e. <0.01 ml/cm^2/day) [37]. However, it should be noted that these rates may depend strongly on the location of the implant, with certain applications (such as stents) allowing increased rates of H_2 evolution due to blood flow. There does not appear to be an absolute rate requirement, and each alloy must be investigated in relation to its intended function

examining this aspect has been taken recently via computational modelling [31]. Mg biomaterials will only ever become acceptable for a variety of biomedical roles through carefully planned and executed, holistic, systematic investigation. In the subsequent chapters, we demonstrate the critical roles of electrochemistry and metallurgy, along with clinical considerations. To make good progress in an area which covers such widely varying fields requires coordination between disciplines—engineers need to have some understanding of medicine, and clinicians need to understand materials science.

1.4.1 Understanding Biocorrosion Experiments and Their Results

The most widely investigated property of potential Mg alloy biomaterials is the "corrosion rate" (as determined from a variety of methods). As discussed in Chap. 2, understanding the corrosion rate that is measured and the mechanisms which affect the corrosion rate are important steps towards its control. Management of corrosion (be it by alloy selection, implant modification or coatings) is necessary to allow appropriate time for healing to occur, to prevent toxicity and related inflammation, to provide sufficient mechanical properties and to control hydrogen evolution. In vitro experimentation should provide confirmation of acceptable levels of these aspects as a necessary first step and allow prescreening of potential alloys.

At present, there are no strong correlations between in vitro and in vivo results [30, 32]. However, an analysis of the current literature (including all references in Appendix C) draws attention to the fact that many, if not most, studies have suffered from a lack of understanding of the appropriate uses of the available in vitro experiments. Widespread problems include use of poorly designed test setups, drawing conclusions based on incorrect assumptions and unsatisfactory analysis of data. These issues appear in many publications to this day—and are highlighted further below. Consequently, the current lack of correlation between in vitro and in vivo results can be attributed to poor choice and execution of in vitro tests and does not preclude the development of such an association at a later date.

1.4.2 In Vitro Variables and Their Effect on Biocorrosion

The design process for in vitro experiments focuses on the choice and control of variables which affect every aspect of the results obtained. Variables such as temperature, medium (or simulated body fluid) and pH control can all alter results in ways that completely change conclusions drawn. Looking at the current literature (Appendix C), it is clear that studies may have overlooked or not controlled for important variables and consequently may not have obtained useful data, making comparisons between studies almost impossible.

In this book, we describe how to carry out the various in vitro experiments, how to analyse the data and how to properly control test variables at appropriate values to obtain realistic, reproducible and comparable results. We also show how inappropriate choices of variables affect the corrosion behaviour of different Mg alloys in an in vitro environment. We then discuss how the composition of Mg alloys affects their biocorrosion, as a first step in understanding how one can design an appropriate alloy that is suitable for use inside the human body.

1.4.3 Catalogue of Current Research

As part of a review of the current literature, we have created a catalogue of studies that outlines the parameters reported, including the following:

- Alloys investigated
- Solutions used
- Buffering system
- Coatings or work applied to samples
- Atmosphere control (e.g. 5 % CO_2)
- Temperature
- pH control
- Experimental procedure
- Time period
- Investigation techniques employed
- Analysis techniques (e.g. optical, SEM, EDS, etc.).

Compiled over several years, this catalogue allows important comparisons to be made between similar studies. The majority of the literature reported was accessed via ScienceDirect®, EngineeringVillage®, ProQuest® and ISI Web of Knowledge®. The authors attempted to collect as much relevant literature as possible, and a large number of reference alerts were created to ensure new publications were included up to the publication date of this book.

The catalogue can be found in Appendix C.

References

1. Park JB, Lakes RS (2007) Biomaterials: an introduction, 3rd edn. Springer, New York
2. Rather BD, Hoffman AS et al (1996) Biomaterial science. Academic Press, London
3. Charnley J (1961) Arthroplasty of the hip: a new operation. The Lancet 277(7187):1129–1132
4. Charnley J (1963) Tissue reactions to polytetrafluoroethylene. The Lancet 282(7322):1379

5. Markets and Markets (2011) Global biomaterials market (2009–2014). http://www.market sandmarkets.com/Market-Reports/biomaterials-393.html
6. Navarro M, Michiardi A et al (2008) Biomaterials in orthopaedics. J Roy Soc Interface 5(27):1137–1158
7. Hench LL, Polak JM (2002) Third-generation biomedical materials. Science 295(5557):1014–1017
8. Wedepohl KH (1991) Chemical composition and fractionation of the continental crust. Geol Rundsch 80(2):207–223
9. Weeks ME (1945) Discovery of the elements. J Chem Educ 22(8)
10. Roberts CS (1960) Magnesium and its alloys. Wiley, New York
11. Kramer DA (2008) Magnesium. 2008 minerals yearbook
12. Friedrich HE (2006) Magnesium technology: metallurgy, design data, applications. Springer, Heidelberg
13. Huse EC (1878) A new ligature? Chicago Med J Examiner 37:171–172
14. Lambotte A (1932) L'utilisation du magnesium comme materiel perdu dans l'osteosynthese. Bull Mem Soc Nat Chir 28:1325–1334
15. Verbrugge J (1933) La tolérance du tissu osseux vis-à-vis du magnésium métallique. Presse Med 55:1112–1114
16. Verbrugge J (1937) L'utilisation du magnésium dans le traitement chirurgical des fractures. Bull Mém Soc Nat Cir 59(59):813–823
17. Verbrugge J (1934) Le Matériel métallique résorbable en chirurgie osseuse. La Presse Med 23:460–465
18. Troitskii VV, Tsitrin DN (1944) The resorbing metallic alloy 'Osteosinthezit' as material for fastening broken bone. Khirurgiia 8:41–44
19. Znamenskii MS (1945) Metallic osteosynthesis by means of and apparatus made of resorbing metal. Khirurgiia 12:60–63
20. Staiger MP, Pietak AM et al (2006) Magnesium and its alloys as orthopedic biomaterials: a review. Biomaterials 27(9):1728–1734
21. Saris N-EL, Mervaala E et al (2000) Magnesium: an update on physiological, clinical and analytical aspects. Clin Chim Acta 294(1–2):1–26
22. Song G (2007) Control of biodegradation of biocompatible magnesium alloys. Corros Sci 49(4):1696–1701
23. Seiler HG, Sigel H (1988) Handbook of toxicity of inorganic compounds. Marcel Dekker Inc, New York
24. Leroy J (1926) Necessite du magnesium pour la croissance de la souris. C R Seances Soc Biol 94:431–433
25. Lusk JE, Williams RJP et al (1968) Magnesium and the growth of Escherichia Coli. J Biol Chem 243(10):2618–2624
26. Vormann J (2003) Magnesium: nutrition and metabolism. Mol Aspects Med 24:27–37
27. Sojka JE, Weaver CM (1995) Magnesium supplementation and osteoporosis. Nutr Rev 53(3):71–74
28. Mudali UK, Raj B et al (2003) Corrosion of bio implants. Sadhana Acad Proc Eng Sci 28(3–4):601–637
29. Kirkland NT (2012) Magnesium biomaterials: past, present and future. Corros Eng Sci Technol 47(5):322–328
30. Witte F, Nellesen J et al (2006) In vitro and in vivo corrosion measurements of magnesium alloys. Biomaterials 27(7):1013–1018
31. Grogan JA, O'Brien BJ et al (2011) A corrosion model for bioabsorbable metallic stents. Acta Biomater 7(9):3523–3533
32. Walker J, Shadanbaz S et al (2012) Magnesium alloys: predicting in vivo corrosion with in vitro immersion testing. J Biomed Mater Res B Appl Biomater 100B(4):1134–1141
33. Park JB, Bronzino JD (2003) Biomaterials: principles and applications. CRC Press, London
34. Puleo DA, Bizios R (2009) Biological interactions on materials surfaces: understanding and controlling protein, cell and tissue responses. Springer, London

35. Department TAFST (2006) Magnesium alloys. The American Foundry Society, Schaumburg
36. Richards AM, Coleman NW et al (2010) Bone density and cortical thickness in normal, osteopenic, and osteoporotic sacra. J Osteoporos
37. Avedesian MM, Baker H (1999) Magnesium and magnesium alloys. ASM International, Materials Park, Ohio
38. Emley EF (1966) Principles of magnesium technology. Pergamon Press
39. Kirkland NT, Kolbeinsson I et al (2011) Synthesis and properties of topologically ordered porous magnesium. Mater Sci Eng B 176(20):1666–1672
40. Staiger MP, Kolbeinsson I et al (2010) Synthesis of topologically-ordered open-cell porous magnesium. Mater Lett 64(23):2572–2574
41. Kirkland NT, Kolbeinsson I et al (2009) Processing-property relationships of as-cast magnesium foams with controllable architecture. Int J Mod Phys B 23(6–7):1002–1008
42. Wintermantel E, Suk-Woo H (1998) Biokompatible Werkstoffe und Bauweisen (Biocompatible material and design), vol 2. Springer
43. Pietrzak WS, Sarver D et al (1996) Bioresorbable implants: practical considerations. Bone 19(1):S109–S119
44. Vadapalli S, Sairyo K et al (2006) Biomechanical rationale for using polyetheretherketone (peek) spacers for lumbar interbody fusion: a finite element study. Spine 31(26):E992–E998
45. Tsantrizos A, Baramki HG et al (2000) Segmental stability and compressive strength of posterior lumbar interbody fusion implants. Spine 25(15):1899–1907
46. Rashmir-Raven AM, Richardson DC et al (1995) The response of cancellous and cortical canine bone to hydroxylapatite-coated and uncoated titanium rods. J Appl Biomater 6(4):237–242
47. Blokhuis TJ, Termaat M et al (2007) Properties of calcium phosphate ceramics in relation for their in vivo behaviour. J Trauma Inj Infect Crit Care 48
48. Allan B (1999) Closer to nature: new biomaterials and tissue engineering. Br J Opthalmology 83:1235–1240
49. Merck A (2006) International water, electrolyte mineral, and acid/base metabolism. In: Porter RS, Kaplan JL (eds) Merck manual of diagnosis and therapy. Merck & Co., Inc
50. Okuma T (2001) Magnesium and bone strength. Nutrition 17:679–680
51. Wolf FI, Cittadini A (2003) Chemistry and biochemistry of magnesium. Mol Aspects Med 24:3–9
52. Hartwig A (2001) Role of magnesium in genomic stability. Mutat Res Fundam Mol Mech Mutagenesis 475:113–121
53. Howlett CR, Zreiqat H et al (1994) The effect of magnesium ion implantation into alumina upon the adhesion of human bone derived cells. J Mater Sci Mater Med 9:715
54. Lopez HY, Cortes DA et al (2006) In vitro bioactivity assessment of metallic magnesium. Key Eng Mater 309–311:453–456
55. Kim SR, Lee JH et al (2003) Synthesis of Si, Mg substituted Hydroxyapatites and their sintering behaviors. Biomaterials 24(8):1389–1398
56. Puleo DA, Huh WW (1995) Acute toxicity of metal ions in cultures of osteogenic cells derived from bone marrow stromal cells. J Appl Biomater 6(2):109–116
57. Granchi D, Ciapetti G et al (1999) Cytokine release in mononuclear cells of patients with Co-Cr hip prosthesis. Biomaterials 20(12):1079–1086
58. Pholer OEM (1986) Failure of orthopaedic metallic implants. In: ASM handbook on failure analysis and prevention, vol 11, 9th edn. ASM International, Metals Park, Ohio, p 670
59. Bach FW (2006) Development of biocompatible magnesium alloys and investigation of the degradation behaviour. In: Sustainable bioresorbable and permanent implants of metallic and ceramic materials. Medical University of Hanover
60. Davis KG, Marras WS et al (1998) Evaluation of spinal loading during lowering and lifting. Clin Biomech 13(3):141–152
61. Polmear IJ (1999) Magnesium and magnesium alloys. In: Avedesian MM, Baker H (eds) ASM specialty handbook. USA, pp 12–25

62. Inoue H, Sugahara K et al (2002) Corrosion rate of magnesium and its alloys in buffered chloride solutions. Corros Sci 44(3):603–610
63. Witte F, Hort N et al (2008) Degradable biomaterials based on magnesium corrosion. Curr Opin Solid State Mater Sci 12(5–6):63–72
64. Seal CK, Vince K et al (2009) Biodegradable surgical implants based on magnesium alloys: a review of current research. In: IOP conference series: materials science and engineering: 012011
65. Meyer-Lindenberg A, Windhugen H et al US 200410241036
66. Zeng R, Dietzel W et al (2008) Progress and challenge for magnesium alloys as biomaterials. Adv Eng Mater 10(8):B3–B14
67. Williams D (2006) New interests in magnesium. Med Device Technol 17(3):9–10

Chapter 2
Magnesium Biocorrosion Experiments

Abstract To obtain useful results regarding the in vitro performance of Mg biomaterials, correct use of the available techniques is vital. The advantages and limitations of the various in vitro techniques must be understood so that appropriate techniques can be chosen and reasonable conclusions drawn. This chapter outlines the most common techniques, the information they yield, their benefits and potential drawbacks.

Keywords In vitro · Electrochemical testing · Mass loss · Hydrogen evolution · Potentiodynamic polarisation · Impedance spectroscopy · Biocorrosion · Degradation · In vivo

2.1 In Vivo Testing

2.1.1 Review of In Vivo Experiments Performed to Date

Since its first recorded use by Aristotle in the fourth century BC, animal research has produced significant advances in medicine [1–3], and it is clear that the benefits obtained from selected in vivo experiments have been and will continue to be immense [4]. Any potential new biomaterial must be tested in applicable animal models before they can be approved for safe use in humans, in line with the various standards that have been set up for this purpose. These are mandatory for assessment of degradation and toxicity.

Of over 30 in vivo studies that have been performed on Mg alloys, the vast majority have used implantation into the femurs of rabbits or guinea pigs. AZ alloys (Mg-Al-Zn) have been the most commonly investigated, with 13 different experiments carried out in rabbits, guinea pigs and sheep. Mg-Ca (nine studies to date) and two Mg-rare earth (RE) alloys, LAE442 (eight studies) and WE43 (four studies), have also been investigated. Other alloys that have been researched include various Mg-Zn-Mn compositions (four studies), AE21 (a single study), Mg-Zn (also one study), pure Mg (one study) and two proprietary alloys. These studies exclude tests carried out in the early 1900s, but do incorporate the few reported investigations of Mg implants in humans. A summary of some of the key findings of each of these tests is shown in Table 2.1 along with the alloy, animal model and implant location.

The majority of the studies in Table 2.1 brought to light potential problems arising from corrosion or by-products (i.e. H_2 gas) of the investigated alloys. These include the need for a greater understanding of the corrosion mechanisms that are occurring in vivo, as well as more systematic studies of individual alloys. However, Mg alloys appear to perform well in vivo and warrant further study.

2.1.2 In Vitro/In Vivo Relationship Not Yet Established

There are only a limited number of studies that have systematically compared Mg alloys both in vivo and in vitro [14, 16, 26, 30, 31, 34, 41]. The majority of results have shown significant differences between corrosion rates in the different environments; however, no report has provided firm reasons for such differences. All of these studies also made at least one "variable mistake", using either a very simple SBF (which is unrealistic compared with the body), or not reporting values of important parameters (such as pH or temperature). This issue, which is discussed further in Chap. 3, limits the use of the results reported and conclusions drawn. A summary of the set-ups used and findings of these correlation studies can be seen in Tables 2.2 and 2.3. It is established in the biomaterial field that in vitro tests will likely never be able to fully emulate every complexity of the human body. The multifaceted reactions that occur between the numerous compounds, proteins and amino acids, would be nearly impossible to recreate outside of a living organism. However, in vitro tests have been developed and widely employed to investigate many current biomaterials, such as titanium, with relative success [42]. These tests have allowed approval for systematic in vivo studies and eventual use in humans. In his book on the biological performance of materials, Jonathan Black stated that "materials must be tested in vitro before implantation, even in animals" (p. 15, [43]). He regarded replicating and testing in the environment that the material will encounter after implantation as crucial to the success of any biomaterial.

The real benefit of in vitro tests lies in the effective screening out unsuitable materials, thus dramatically reducing the number of animal and human trials. Non-animal experiments also offer a number of benefits including reduced cost, controllability, ease of monitoring/recording of data and reproducibility. Some of the primary drawbacks of in vivo experiments are summarised in Table 2.4.

Table 2.1 In vivo tests on Mg alloys

Primary author (references)	Main findings	Material/animal/ location
Celarek [5]	• ZX50 degraded too quickly, WZ21 was suitable. BMG is promising and lacks mechanical properties	ZX50, WZ21MgZnCa BMG/rats/ femora
Chen [6]	• Rods coated with HA degraded slowly, showing newly formed tissue—indicating a good response	MgZnCa/rabbits/ femora
Duygulu [7]	• Mg alloys have "significant potential" as implant materials	AZ31/sheep/hips
Erdmann [8]	• Lost mechanical integrity after 6 weeks • Initial (3 week) ability similar to S316L steel	Mg0.8Ca/rabbits/ tibiae
Fischerauer [9]	• Implants displayed a positive bone response and could be suitable for certain applications	ZX50/rabbits/ femora
Heublein [10]	• AE21 is a "realistic alternative to permanent implants"	AE21/pigs/stents
Krause [11]	• Mg0.8Ca of limited use, WE43 corrosion is too non-uniform, LAE442 shows good promise	Mg0.8Ca, WE43, LAE442/rabbits/ tibiae
Kraus [12]	• WZ21 did not lose integrity; both alloys were deemed suitable for future investigation	ZX50, WZ21/rats/ femora
Krause [13]	• Mg0.8Ca promising, WE43 fracture strength is unpredictable; LAE442 has good corrosion/ strength potential	Mg0.8Ca, WE43, LAE442/rabbits/ tibiae
Li [14]	• Alloys showed very good promise as potential implants; secondary phases may be used for control	MgCa(1,2,3)/rabbits/ femora
Reifenrath [15]	• AZ91 cannot sufficiently replace subchondral bone during the first 12 weeks (of implantation); appropriate coatings may slow the corrosion	AZ91/rabbits/knees
Ren [16]	• Alloy displayed appropriate degradation rate and low H_2 release, formed CaP similar to natural bone	AZ31B/rabbits/ femora
Thomopoulos [17]	• Mg-based bone adhesive (MBA) displayed a negative reaction, perhaps due to "allergic response"	MBA/dog/distal phalange
Von Der Hoh [18]	• "In summary, it can be said that all magnesium implants investigated were well tolerated"	Mg0.8Ca/rabbits/ femora
Von Der Hoh [19]	• All implants tolerated well, with no redness or excess pain, early wound healing, though callus formation was observed around implants	MgCa (0.4–2 %)/ rabbits/femora
Waksman [20]	• Stents were safe over this period (28 days), but longer-term tests are needed	WE43/pig/Artery
Waksman [21]	• WE43 performed well as stent, although incomplete healing was observed	WE43/pig/artery
Witte [22]	• No allergenic reactions/ skin sensitising potential was observed in skin biopsies	AZ31, AZ91, WE43, LAE442/guinea pigs/intradermal

(continued)

Table 2.1 (continued)

Primary author (references)	Main findings	Material/animal/ location
Witte [23]	• "Both alloys showed direct contact with newly formed bone," proves biocompatibility	AZ91D, LAE442/ guinea pigs/ femora
Witte [24]	• "All magnesium implants have been observed in direct bone contact and without a fibrous capsule"	LAE442/rabbits/ femora
Witte [25]	• High mineral apposition rates and increased bone mass were observed, no bone in soft tissue, "high magnesium ion could lead to bone cell activation"	AZ31, AZ91, WE43, LAE442/guinea pigs/femora
Witte [26]	• "Alloys had direct contact with bone which showed good compatibility"	AZ91D, LAE442/ guinea pigs/ femora
Witte [27]	• Alloys degraded too rapidly in vivo to allow cartilage repair; however, cartilage tissue was not negatively affected; new bone was observed	AZ91/rabbits/knee
Witte [28]	• Extended peri-implant bone remodelling with good biocompatibility	AZ91D/rabbits/knee
Witte [29]	• "Promising" results for use of Mg as replacement for subchondral bone	AZ91D/rabbits/knee
Wong [30]	• New bone formation was observed, no inflammation, necrosis or accumulated hydrogen gas	AZ91/rabbits/ trochanters
Xu [31]	• CaP-coated Mg alloy had good surface bioactivity and promoted early bone growth	Mg-1.2Mn-1Zn/ rabbits/femora
Xu [32]	• After 18 weeks, all implants were fixed and no inflammation was visible, Mn and Zn distributed homogenously, easily adsorbed	Mg-1.2Mn-1Zn/rats/ femora
Zhang [33]	• New bone in tight contact with implant due to good osteoconductivity, increased bone around implant, no increase in fibrous membrane	Mg-1Zn-0.8Mn/rats/ femora
Zhang [34]	• No harm to important organs or adverse effects of the generated hydrogen nor released Zn	Mg-6Zn/rabbits/ femora
Human trials		
Bosiers [35]	• Safe for treating peripheral artery disease, but not for long-term use yet	AMS proprietary/ adults/stent
Erbel [36]	• Safely degraded over 4 months • Similar effects as other metal stents	AMS proprietary/N/ S/stent
McMahon [37]	• Initial increase in vessel diameter was followed (4 months) by significant restenosis	AMS proprietary/ baby/stent
Peeters [38]	• Adequate clinical performance, suitable for CLI patients	AMS proprietary/ adults/stent
Schranz [39]	• Effectively used to treat aortic coarctation for the newborn	AMS proprietary/ baby/stent
Zarter [40]	• Well tolerated, serum levels were only moderately elevated	AMS proprietary/ baby/stent

Table 2.2 Comparison of corrosion parameters from studies with in vitro and in vivo tests

Primary author (references)	Alloy(s)	In vitro		In vivo
		Sample area (cm^2)/solution amount/type	Corrosion rates (mm/year)	Corrosion rates (mm/year)
Witte [26]	AZ91D LAE442	3.4/25 l/artificial sea water	PDP = 2.8, ML = −0.267 PDP = 6.9, ML = 5.535	SRµCT = 3.5 × 10^{-4} SRµCT = 1.21 × 10^{-4}
Zhang [34]	HP Mg Mg-6Zn	1/N/S/basic SBF	PDP = 0.2, ML = 0.1–0.4 PDP = 0.16, ML = 0.07–0.2	ML = 2.32
Li [14]	Mg-1Ca Mg-2Ca Mg-3Ca	2.2/N/S/Kokubo's SBF	PDP = 12.56 PDP = 12.98 PDP = 25.00	ML = 1.27[a]
Wong [30]	AZ91	1/N/S/basic SBF	ML = 5.9 × 10^{-6}	ML = 1.2 × 10^{-10b}
Ren [16]	AZ31B	9.6/N/S/HBSS	ML = 0.2–0.7	–
Ren [41]	HP Mg	11.8/N/S/HBSS	ML = 0.2–0.6	–
Xu [31]	Mg-1.2Mn-1Zn	1.57/N/S/ PRI1640	–	–

N/S The values were not stated in the work
[a] Degradation rate converted from reported mg/mm^2 /y
[b] Converted to mm/year using ASTM G31 [44]

2.1.3 Classification of In Vitro Tests

In vitro tests for Mg are classified into two broad categories, tests for: (1) bio-corrosion rates (or biodegradation behaviour) or (2) toxicity/interaction with biological organisms. These aspects are related, since rapid degradation can lead to toxicity and other negative biological reactions. Toxicity is almost always assessed in the presence of living cells, while corrosion testing does not have this requirement. Corrosion of Mg alloys is a complex process, requiring a combination of different techniques for complete characterisation [49]. The various test methods available to measure corrosion are grouped into two categories: (1) un-polarised and (2) polarised. The difference between these methods relates to the presence of a driving force (i.e. electrochemical polarisation) which is applied or measured during the test.

There are several commonly used in vitro tests, from simple immersion or mass loss experiments to electrochemical investigations, which provide mechanistic information on how the material degrades. The various in vitro methods have their own advantages and limitations. Inappropriate or incomplete testing and inter-pretation of the results may lead to unnecessary in vivo experiments, performed on materials that are likely to fail. It is therefore crucial to fully understand the correct uses of each experiment, and perhaps more importantly, the shortcomings which need to be considered before drawing conclusions.

Table 2.3 Summary of studies that performed in vitro/in vivo experiments concurrently

Author (references)	Study overview	Key findings
Quantitative studies		
Witte [26]	• pH = 8.2 (not controlled), T = N/S •Gain in mass loss. Mass loss due to unremoved corrosion products • Authors since rectified issues [45, 46]	• No correlation between in vitro and in vivo; artificial sea water is an inappropriate SBF because of lack of inorganic salts
Zhang [34]	• pH = 7.44 (not controlled), T = 37 °C • pH rose to 9 after 24 h (unrealistic)	• Electrochemical and immersion corrosion rates were similar; in vivo rates were up to 40× faster
Li [14]	• pH = 7.4 (not controlled), T = 37 °C • pH rose to 12 after 24 h (unrealistic), solution based on ASTMASTM G31 [44] (not appropriate)	• In vivo only performed on Mg-1Ca; in vitro displayed 10× faster rate of corrosion, minimal discussion of reasons for corrosion rate
Wong [30]	• pH = 7.4 (not controlled), T = 37 °C • 2 months with no pH control, quickly rose to above 8 for some samples. • μCT limited use due to problems discerning between corrosion product and implant. • No report of ML or Tafel analysis	• Mg ion loss, converted to mass loss, was extremely small; barely noticeable (through μCT) amounts of corrosion occurred in vivo
Qualitative studies		
Ren [16]	• pH = 7.5 (controlled), T = 37 °C • Unknown solution amount, only in vitro corrosion rates reported	• In vitro samples displayed "acceptable" corrosion rates (according to author), in vivo alloy showed good CaP build-up around implant
Ren [41]	• pH = 7.5 (controlled), T = 37 °C • Unknown solution amount, only in vitro corrosion rates reported	• 10× increase in corrosion rate (in vitro) if pH was not controlled, in vivo Mg displaced good thromboresistant properties
Xu [31]	• Tests primarily used to determine toxicity effects of alloy in vitro and in vivo	• CaP coatings on Mg resulted in good surface bioactivity (in vitro) and early bone growth (in vivo)

The terms "biocorrosion" or "biodegradation" are used synonymously in this work to refer to corrosion that takes place in a simulated body environment. Unless otherwise stated, all experiments by the authors reported in this book were performed at physiological temperatures (37 ± 0.5 °C), pH (7.4 ± 0.1) in Hank's balanced salt solution buffered with HEPES.

Table 2.4 Disadvantages of in vivo tests

Disadvantage	Description
Discomfort to test animals	• If not properly tested beforehand, material may cause significant pain/discomfort to the animal
	• Majority of countries worldwide have strict policies regarding requirements before in vivo tests may be performed
Cost	• Intrinsically high cost to perform
	• Multiple samples for statistical accuracy
	• Failed experiments more common due to uncontrollable factors/variations
	• No simple cost-to-benefit ratio exists worldwide [47]
Time	• Experiments often cannot commence for long periods, while approval is obtained from regulatory body
Difficulty to analyse in situ	• Possible to use X-ray and μCT, but greater qualitative/quantitative analysis requires animal killing
Choice of animal model	• Especially important in orthopaedic applications, where mechanical conditions play role in biomaterial performance (e.g. rabbit may not be suitable for testing spinal implant materials)
	• For cartilage repair, no current animal species replicates all of the bioproperties [48]; extrapolation across species is difficult
	• Choice of most appropriate model is often problematic
Ethical issues	• Animal rights is closely followed and a popular issue in modern culture
	• Rationalisation of studies depends on "potential human benefit and justification for their exploitation" [47]

2.2 Physical Testing

2.2.1 Mass Loss

Often referred to as weight loss, a mass loss (ML) measurement is perhaps the simplest in vitro method available for investigating Mg corrosion. Set-up may vary depending on the experimental variables, such as choice of buffering system, but the same design is typically employed, requiring only a sample, solution and an accurate microbalance. During the test, the sample is placed in a selected solution for a set period of time, after which it is removed and ML is measured. Normally, a mixture, such as dilute chromic acid, is used to remove any corrosion product on the surface.

Perhaps in part due to their inherent simplicity, ML experiments have been relatively popular in the literature, with over 40 reported studies in which ML has been utilised to quantify Mg biocorrosion (see Appendix C). Typically, results obtained from ML tests are accurate, assuming corrosion layer removal issues are minimised and corrosion is relatively general. In addition, depending on the set-up design, it is possible to concurrently perform other in vitro tests with ML, such as pH monitoring or electrochemical experiments, although this may inadvertently alter results [50].

2.2.1.1 Corrosion Mechanisms Not Revealed

ML experiments require a relatively extensive degree of corrosion to take place for precise measurement of the mass change, and multiple replicates are needed. Additionally, while ML experiments reveal the physical amount of corrosion that has taken place, they do not divulge the mechanisms involved in the corrosion process. For example, although it may be perceived that a given alloy corrodes faster than another, ML does not provide the information required to determine why this happens, as evidenced by Fig. 2.1. Here, it can be seen that similar total mass loss values were obtained for two different alloys after one week of immersion; however, comparison of the polarisation curves indicated the alloys had different corrosion potentials and relative anodic and cathodic reaction kinetics. This behaviour is particularly important in more physiologically realistic media, as the organic components, such as amino acids and proteins, have diverse effects on each partial reaction.

2.2.1.2 Sample Surface Area to Media Ratio

The corrosion of Mg (or any metal) will be heavily influenced by a change in pH. The corrosion rate of Mg will monotonically decrease as a function of increasing pH, with passivation attained at a pH of >11 [51]. It is known that the corrosion of Mg will result in formation of Mg^{2+} and reduction of H_2O, releasing H_2 gas and hydroxyl (OH^-) ions into the surrounding medium. Consequently, a measured

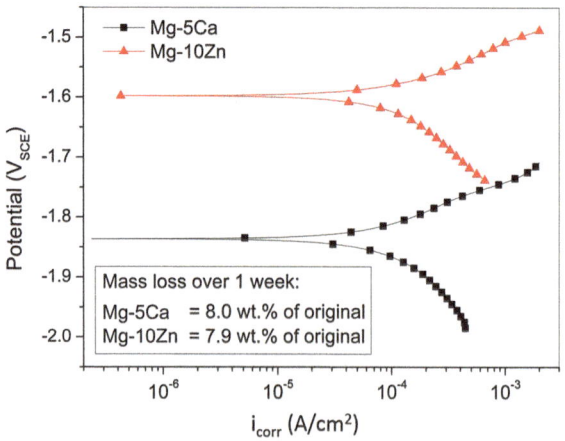

Fig. 2.1 Polarisation responses and mass losses of Mg-5Ca and Mg-10Zn after 1 week

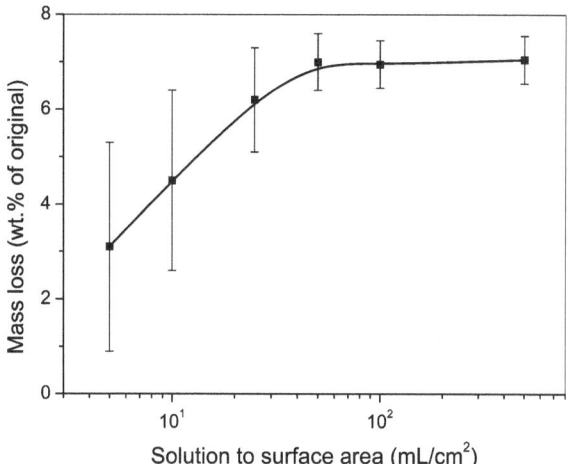

Fig. 2.2 Mass loss as a function of solution to surface area for pure Mg after 1 week

corrosion rate may be heavily affected by shifts in the pH level because of production of OH^-, and these changes depend upon the ratio of volume of medium to sample surface. To investigate this effect, the ratio was varied from 5 to 500 ml/cm^2, ML tests of pure Mg over 1 week (Fig. 2.2). In this case, the pH was adjusted every 30 min to 7.4. Mass loss rose steadily as the medium volume to area ratio increased, reaching a plateau at a ratio of 50 ml/cm^2. This rise is probably because of rapid increases in the local pH at the sample surface that occur in between each readjustment of the pH and decrease the potential for corrosion [52]. Moreover, an increase in pH is known to encourage the formation of $Mg(OH)_2$ and CaP layers on the Mg substrate [53, 54], providing additional protection of the underlying surface from the corrosive media.

Standards outlined by the American Society for Testing and Materials (ASTM) for determining ML of structural metals, which have been referenced in many Mg biomaterial studies, are not suitable for Mg, since the suggested solution volumes (0.2–0.4 ml/mm^2) are too small to accommodate the rapid change in local pH that typically accompanies Mg corrosion [26, 44] at T_{phys}. In the absence of standardised electrolyte volume guidelines, authors in the wider literature have employed a widespread range of medium volume to sample area ratios, making meaningful comparisons across different studies all but impossible. One step in the right direction would be to base the sample medium volume to surface area ratio on what is clinically relevant, such as a surface area calculated for a specific implant and the available total serum in the human body [55]. However, current best practice would be to maintain the volume of medium as large as practicably possible; the results shown earlier suggest that the volume should be at least 50 ml/cm^2 (Fig. 2.2).

2.2.1.3 Corrosion Product Removal

With ML experiments, it is important to consider the removal of corrosion product after immersion. Many early Mg in vitro studies failed to take this step into account, with some studies even reporting a net gain in mass due to the hydrated corrosion product layer on the surface [26, 56]. The hydrated corrosion layer on Mg has no mechanical integrity, and therefore, its inclusion is not useful in an engineering context. Normally, a mixture of $AgNO_3$ and CrO_3 is used to effectively remove the corrosion layer by dissolving $Mg(OH)_2$. This cleaning process causes negligible corrosion to the underlying Mg surface [57–59].

2.2.2 Hydrogen Evolution Measurement

The corrosion reaction for Mg in a neutral aqueous environment is

$$Mg(s) + 2H_2O(aq) = Mq^{2+}(aq) + 2OH^-(aq) + H_2(g) \tag{2.1}$$

This indicates that one atom of Mg will generate one hydrogen gas molecule, or, in other words, the evolution of one mol of hydrogen gas (22.4 l) directly corresponds to the dissolution of one mol of Mg (24.31 g). Consequently, in theory, measuring the volume of H_2 gas produced is equivalent to measuring the Mg ML.

The basic set-up of a hydrogen evolution (H_2^{evo}) experiment is similar to that of a standard ML experiment. Typically, a sample is submerged in a solution, and a "collector" is placed above. For most set-ups, this involves the use of an inverted funnel and burette or graduated tube [53, 60–62]. Before the investigation starts, the burette is filled with test solution, and, as the H_2 gas evolves, it will rise and replace the solution in the burette. In the literature, H_2^{evo} experiments have been used in over 35 studies to determine Mg in vitro performance (see Appendix C).

2.2.2.1 Key Benefits of Hydrogen Evolution Measurement

Unlike ML experiments, collection of H_2^{evo} data is unaffected by the formation of corrosion products, assuming such products remain within the capture region of the collector. The principal advantage of H_2^{evo} is its ability to take measurements at multiple time points throughout the test, which allows analysis of time-dependant changes in corrosion rate—information not easily obtainable via ML experiments. To give an example of this, imagine two different Mg alloys, X and Y. While a ML test will only show that alloy X has corroded less than alloy Y, results from H_2^{evo} would reveal that alloy X is degrading rapidly towards the end of experiment lifetime, likely exhibiting increased corrosion at longer immersion times, compared with alloy Y.

Based on Eq. 2.1, H_2^{evo} experiments permit the possible determination of the degree of alkalisation based on the volume of hydrogen evolved, assuming that two mol of $(OH)^-$ is released for every one mol of Mg that is oxidised. However, this does not take into account the establishment of corrosion layers and other products that may form in the solution, and therefore, care must be taken when using such an extrapolation.

2.2.2.2 Drawbacks of H_2^{evo}

Similar to ML, the H_2^{evo} technique will not provide information on the corrosion mechanisms that are occurring at the sample surface, so the reasons for changes in evolution rate are not apparent from the test. Additionally, there may be an insufficient volume of evolved gas to allow accurate and reproducible measurements until significant corrosion has occurred. This means that the H_2^{evo} technique is generally not suitable for either (1) study of the earliest stages of corrosion (<1 h after immersion) or (2) highly corrosion resistant alloys that liberate low levels of H_2.

The H_2^{evo} technique also does not allow straightforward investigation of the effect of flow rates on Mg corrosion, as evolved H_2 gas would be difficult to capture accurately. Although possible, it is also increasingly difficult to set up concurrent electrochemical tests when compared with ML experiments, because the extra glassware (burette, funnel) makes electrode placement challenging.

2.2.2.3 Considerations When Performing H_2^{evo} Experiments

There are a number of considerations and potential sources of error that must be taken into account when setting up, performing and analysing H_2^{evo} experiments. Gas solubility variation with temperature [63] needs to be considered for H_2^{evo} experiments, or where temperature variation takes place. For 1 l of water, a total of 2.14 ml and 4 ml of O_2 and N_2, respectively, would be released when the temperature changes from 20 to 37 °C (Table 2.5). However, although important, these measurements are for a larger volume than is normally used, and assume that all the released gas is captured by the H_2^{evo} apparatus. Although it would appear unlikely that such a release of gas would have a considerable effect on the overall recorded H_2 level, this factor should be kept in mind during experimental design stages.

Table 2.5 Change in solubilities of O_2 and N_2 and potential release of gas for 1 l of H_2O at pH 7.0 at 1 atm

Gas	Solubility at 20 °C (mg/l)	Solubility at 37 °C (mg/l)	Gas density (mg/ml)	Release of gas (ml)
O_2	9	6.2	1.31	2.14
N_2	18	13	1.25	4

Data from [63]

The evolution of hydrogen is the cathodic reaction of Mg corrosion. Corrosion mechanisms such as dealloying or particle undermining (i.e. secondary phase particles released from the sample) will not be accurately detected by the H_2^{evo} method, resulting in underestimates of corrosion rate, particularly when testing heavily alloyed Mg alloys. Hydrogen evolution will also not correlate with a real corrosion rate in cases where the water reduction reaction is not the dominant or sole cathodic reaction (e.g. Mg-based metallic glasses).

Based on a typical H_2^{evo} experiment using 250 ml (0.25 kg) of water (i.e. SBF), a total of 3.88 ml of H_2 gas is dissolved within this solution. This estimate is based on the solubility of hydrogen gas in H_2O at 37 °C (1.4 mg/kg [64]) and density of hydrogen gas at this temperature ($\sim 9 \times 10^{-5}$ g/ml [65]). Consequently, depending on a number of factors, including medium aeration and preparation steps, the dissolved hydrogen is a significant potential source of error in the measured hydrogen evolution. This is especially important because (1) most reported experiments have measured less than 4 ml of H_2 gas over the entire course of the test, and (2) the precision of a typical burette used for this purpose may only be ±0.1 ml. Therefore, careful control of media and following of strict media preparation steps are necessary to minimise this effect (control experiments should also be performed to see if gas is produced without a sample).

The ideal gas law, which governs the volume of gas produced per mol of H_2, is

$$PV = nRT \tag{2.2}$$

where P is the pressure (Pa), V is volume (m^3), n is the amount of the gas in moles, R is the ideal gas constant, and T is the temperature (K).

Given that η, R and T are controlled in a normal in vitro set-up, the volume of evolved hydrogen will be related solely to the surrounding atmospheric pressure. Consequently, variations resulting from different atmospheric pressures may be significant. For example, experiments performed at an elevation of 2,000 m, an altitude similar to that of Mexico City, would evolve almost 30 % more volume of H_2 than those performed at sea level (such as in Melbourne). Accordingly, all calculations using H_2 measurements should take atmospheric pressure into account using:

$$\Delta W = 1.085 V_u / P_{ATM} \tag{2.3}$$

where ΔW is the change in mass (mg), V_H is the volume of hydrogen that has evolved (ml), and P_{ATM} is the atmospheric pressure (atm).

Even the materials used to perform H_2^{evo} experiments need to be carefully selected. Hydrogen gas is known to be permeable through many polymeric materials [66]. The diffusion of hydrogen through such materials must be considered when designing a suitable test set-up, as choice of inappropriate materials will alter results in uncontrollable ways. H_2 has been found to leak quickly when stored in plastic bottles and may completely disperse within 2 weeks [67]. Consequently, it is recommended that all equipment used to capture the released gas should be made of glass, and extra care should be taken to avoid any leaks.

Fig. 2.3 Photographs (2×) of the *bottom* surfaces of pure Mg samples **a** *before* and **b** *after* immersion for 72 h

In H_2^{evo} experiments published in the literature, whether the underside of a test sample (the part in contact with the bottom of the beaker or mount) is covered to prevent corrosion is rarely reported. For calculation of the H_2 gas evolution rate, which is based on the surface area exposed to the medium, this is an important step in the preparation of the sample, as an uncoated underside will partially corrode, as shown in Fig. 2.3. All surfaces that are not meant to be included in measurements should be covered with an appropriate sealant to prevent their inadvertent degradation.

2.2.2.4 Lack of Correlation in the Literature

Perhaps the most alarming factor facing the acceptance of H_2^{evo} results is the fact that many of the studies in the literature which have compared the total amount of H_2 (converted to equivalent mass loss) with actual ML have not found agreement between the values (i.e. a 1:1 ratio) (Table 2.6). Several studies have found this ratio to vary between 0.22 and 0.323 [16, 41, 68], while others found a closer correlation of 0.89–1.31 [59–61, 69]. It should be noted that all but one of the studies that found a ratio close to 1:1 were performed by the same research groups. This apparent disparity between the theoretical and actual results was also established by the present authors, who found a ratio of ~0.6, even after considering all of the above factors when performing the H_2^{evo} experiments.

The reasons for the lack of strong correlation between captured H_2 and actual mass loss remain unclear. The collection of H_2, although a useful technique, is very susceptible to error and is not efficient. Unaccounted for H_2 may be forming as bubbles on the side of the funnel, getting trapped beneath the specimen or diffusing into the solution. These experimental challenges emphasise the importance of performing other corrosion experiments in combination with H_2^{evo} to

Table 2.6 Mg studies where H_2^{evo} and ML experiments have been concurrently performed

Alloy	Solution	Temp/pH/time (h)	H_2 equivalent ML (mg/cm^2)	Measured ML (mg/cm^2)	ML ratio (H_2 equivalent : ML measured)	References
Pure Mg	1 M NaCl	Rm/N/S/48	0.44	0.43	1.023	[70]
AZ91D	1 M NaCl	Rm/N/S/48	3.1	3.0	1.03	[70]
ZE41	1 M NaCl	Rm/N/S/48	5.9	5.7	1.03	[70]
Mg-0.8Ca	0.9 % NaCl	N/S/N/S/360	0.10	0.31	0.323	[68]
Mg-2Gd	1 % NaCl	21.5/6.5/N/S	6.4	6.4	1	[59]
Mg-5Gd	1 % NaCl	21.5/6.5/N/S	1.97	1.78	1.11	[59]
Mg-10Gd	1 % NaCl	21.5/6.5/N/S	0.49	0.54	0.91	[59]
Mg-15Gd	1 % NaCl	21.5/6.5/N/S	10.95	8.38	1.31	[59]
AZ31B	HBSS	37/7.5/720	0.038	0.14	0.27	[16]
Pure Mg	HBSS	37/7.5/168	0.021	0.094	0.22	[41]
MEZr	5 % NaCl	N/S/N/S/24	25	28	0.89	[69]
99.96 Mg	1 M NaCl	N/S/6/N/S	50*	47	1.06	[60]
99.96 Mg	1 M NaCl	N/S/11/N/S	20*	19.5	1.03	[60]
99.96 Mg	1 M NaOH	N/S/N/S/N/S	0.07*	0.07	1	[60]
99.96 Mg	1 M HCl	N/S/N/S/N/S	~70,000*	~70,000	1	[60]
99.96 Mg	5 % NaCl	Rm/N/S/N/S	20*	20	1	[60]
A6	1M NaCl	N/S/N/S/N/S	35*	37	0.95	[60]
AZ21	1M NaCl	N/S/N/S/N/S	0.50*	0.47	1.06	[60]
AZ91D	1M NaCl	N/S / N/S / N/S	0.2*	0.205	0.98	[60]
AZ91D	5% NaCl	N/S / N/S / N/S	0.21*	0.21	1	[60]
AZ91	SBF	N/S / N/S / 168	0.43*	0.47	0.915	[61]
Data from our work						
Pure Mg	HBSS	37/7.4/72	5.47	9.1	0.601	
Mg-1Ca	HBSS	37/7.4/72	5.40	8.3	0.651	
Mg-1Zn	HBSS	37/7.4/72	3.35	6.2	0.540	

N/S indicates that the value was not stated, and Rm indicates room temperature

*The values were converted from the provided data to mg/cm^2/day or were obtained from graphs

Data presented for this work are the average of 3 samples for each data point

confirm any results obtained. Further discussion of these phenomena can be found in the literature [71].

2.3 Electrochemical Testing

2.3.1 Potentiodynamic Polarisation

Potentiodynamic polarisation (PDP) is currently the most commonly used electrochemical technique for studying in vitro corrosion of Mg alloys, with over 70 studies reported (see Appendix C).

Typically, the test starts with a set period of time where the open circuit potential (OCP) is recorded. This allows the material surface to "stabilise" with the electrolyte and reach a near-steady potential. Following this, an applied potential is imposed between the working and counter electrode. Potential is altered at a controlled rate (e.g. 1 mV/s), and the current required for this change to occur is recorded. Normally for Mg, the potential is initially set to be more negative (cathodic) and sweeps to become more positive (anodic) than the OCP. This limits the effect of the increased Mg dissolution that occurs in the anodic region.

2.3.1.1 Primary Benefits of PDP

PDP sample preparation is a relatively straightforward process, requiring only consistency on behalf of the investigator, especially if a flat cell is employed. A typical single scan for Mg can take as little as 5 min to complete, meaning many scans may be performed in a relatively short time period. As the test is only destructive to the surface being tested (and normally only ~ 50-100 μm deep), a sample may be tested as many times as required, requiring only removal of the damaged layer through regrinding and cleaning between scans. This helps to minimise any variation between samples that is likely to be encountered in ML or H_2^{evo} experiments.

Perhaps the most advantageous aspect of PDP is that it permits quantification of the relative rates of the anodic and cathodic reactions over a wide range of potentials. This is vital in the endeavour to unravel the mechanistic aspects of Mg alloy biocorrosion, allowing an engineering basis for alloy design and selection. This is highlighted in an example of two alloys which possess a similar i_{corr} value but display significantly different anodic and cathodic kinetics. Such an example is clearly shown for pure Mg and Mg-2Zr, where both alloys have a similar i_{corr} (of ~ 30 μA/cm^2) yet displayed significantly different E_{corr} values because of a shift in cathodic kinetics (Fig. 2.4). In this case, the addition of Zr resulted in an increase in cathodic reaction kinetics while simultaneously contributing towards a

Fig. 2.4 Polarisation *curves* for pure Mg and Mg-2Zr. *Arrows* indicate cathodic shift

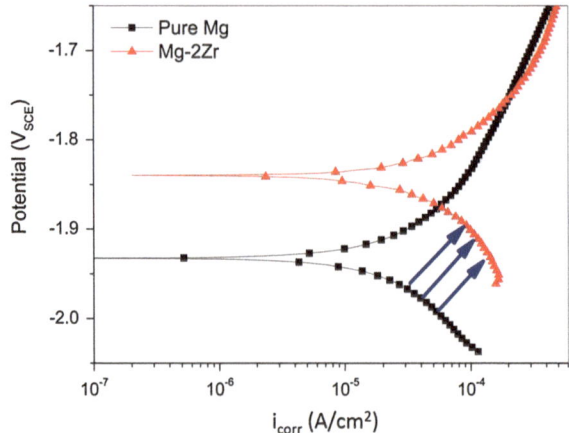

retardation of anodic kinetics over a range of potentials. These underlying reasons for changes in corrosion are crucial to effectively designing alloys for specific corrosion rates.

When combined with microelectrochemical methods, PDP may be used to investigate the detailed effects at the microscale of the various phases in multiphase Mg alloys on biocorrosion in SBF (see [72]). PDP may be employed to determine the contributions of each individual phase to the overall corrosion mechanisms of an alloy; these results may then be used to tailor alloy microstructures for different corrosion rates and applications [73]. In addition, polarisation experiments may be used to determine the effect of organic components, such as amino acids and proteins, on individual phases. This allows optimisation of alloy microstructure for the biological environment, explored for Mg-Ca alloys [74].

2.3.1.2 Drawbacks of PDP

The majority of uncoated Mg alloys will not corrode uniformly. However, the conversion of i_{corr} (expressed as a current) to a corrosion rate (in terms of penetration) requires the assumption that only general corrosion is taking place [75]. Consequently, PDP results cannot typically yield an absolute corrosion rate for Mg, but rather provide an indication of the severity of the corrosion that is taking place at a selected point in time (in terms of current density). At this stage, it should be noted that none of the methods described in this chapter will provide an "absolute" corrosion rate in mm/y unless all the corrosion is assumed to be uniform. Nonetheless, corrosion rate expressed as current density is highly accurate, with the highest absolute precision of the techniques discussed here (although this current may be emanating from a number of local sites on the surface).

The dwell time prior to the commencement of a PDP scan can affect the results. When any metal is initially exposed to an electrolyte, it will take a finite amount of time for the surface to form the electrical double layer (EDL) and for the associated redistribution of species in the electrolyte to take place [76]. Executing scans before this stabilisation period has completed can and will result in inconsistencies in the current required to alter the potential, and consequently, the results will not accurately reflect the electrode reactions [77]. Consequently, best practice would be to define an appropriate waiting time by recording the OCP for each alloy and determining when it becomes stable. However, it remains difficult to know to what extent the surface area may have altered by this point.

The potential scan rate used in PDP has an effect on the amount of current that is required to achieve a potential shift [78]. This parameter must be set by the user before starting experiments. Too slow a scan allows corrosion to influence the surface, while the scan is being performed, especially for highly reactive alloys; too rapid a scan does not allow sufficient time for the system to respond to the changing potential and therefore is dominated by capacitive effects. A number of different scan rates were studied for pure Mg (Fig. 2.5). The slower scan rates (10 and 1 mV/s) resulted in relatively similar curves. However, we have previously found that 10 mV/s is often not suitable for faster corroding alloys, as it leads to significant variations between repeated scans of the same sample. Data from the slowest scan rate tested, 0.1 mV/s, began to show signs of "noise," particularly in the anodic curve. This is probably because the potential is forced away from the OCP for extended periods of time, allowing surface alteration from dissolution to occur.

The parameters used to perform Tafel-type analysis are intricately linked to the results obtained and to the final decision about whether or not an alloy has performed suitably. Because of the logarithmic nature of the current density scale, even small changes in the determined Tafel slopes can result in disparities of up to an order of magnitude in reported i_{corr}. Consequently, analysis of the same data by different researchers can result in quite different, or even completely opposite, conclusions.

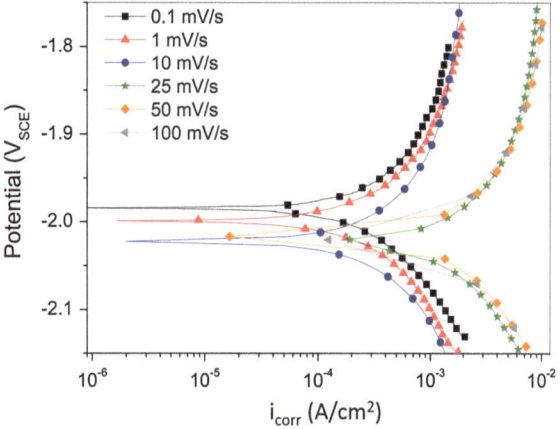

Fig. 2.5 Polarisation *curves* of pure Mg using a range of scan rates from 0.1 to 100 mV/s

It is widely accepted that analysis of polarisation data should not be performed within 50 mV of E_{corr}, as the assumptions that underpin Tafel fits do not apply close to E_{corr} [76]. Nonetheless, it is difficult to set absolute parameters for analysis because of variations between different alloys and solutions; some alloys may even corrode so rapidly during the forced oxidation reaction that only the cathodic slope can be successfully analysed. Hence, reporting of Tafel-type analysis in published works must be accompanied by a description of the potential range over which the analysis was carried out, as well as the software and method used to determine i_{corr}.

PDP is an instantaneous test, representing only a snapshot of the corrosion at the time it is performed. This is analogous to measuring the temperature at one time point compared with getting the average temperature over a day. Although it is understood that the i_{corr} at earlier stages of corrosion is not necessarily representative of all time points, the results are still indicative of the corrosion mechanisms that are taking place and provide a useful tool for comparing multiple variables (e.g. seeing the effects of changing temperature or alloys). By their very nature, polarisation tests will perturb the surface of the sample significantly, due to the large currents achieved over a short period of time. With Mg alloys that corrode relatively quickly, this allows only a single scan to be performed on a sample before it must be removed and repolished. Therefore, cyclic voltammetry, a useful polarisation technique widely used to analyse surface layer formation [79], is unsuitable for Mg alloys in SBF, and PDP cannot typically be used to reveal the individual contributions of layers formed on the surface of Mg. In addition, surface analysis (e.g. scanning electron microscopy) of the surfaces after such polarisation is not particularly relevant, because the products of the artificially accelerated reactions are not necessarily the same as those of normal corrosion (the latter corresponding to a situation where anodes and cathodes coexist on the same surface).

2.3.2 Electrochemical Impedance Spectroscopy

Electrochemical impedance spectroscopy (EIS) employs a low-magnitude AC polarisation, cycling from peak anodic to peak cathodic over a range of frequencies, allowing resistance and capacitance values to be obtained for each frequency. EIS has grown increasingly popular in the field of Mg corrosion in recent years, with over 50 publications having used the technique to study the corrosion of Mg alloys for biomedical applications (See Appendix C)

2.3.2.1 Key Benefits of EIS

EIS provides near-instantaneous information regarding the impedance of a surface subject to minor polarisation [80]. This impedance is inversely proportional to corrosion rate and can be used as an index to the rate at which dissolution is

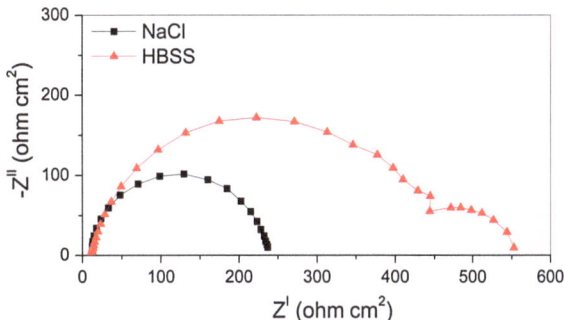

Fig. 2.6 Nyquist plot of pure Mg in a 1.0 wt.% NaCl solution and HBSS after 2 h immersion. Second *semicircle* is indicative of layer formation on surface

occurring [81]. Unlike PDP, EIS can be considered a non-destructive technique when applied to Mg in SBF, allowing for (1) multiple recordings upon the same sample without having to repolish the sample surface after each scan and (2) real-time online monitoring.

Perhaps the greatest advantage of EIS is its ability to detect individual layers on the surface of Mg (or any alloy) [82]. Used over longer periods, it can detect the formation of a corrosion or passivation layer (e.g. CaP) on the surface, as well as revealing how much each layer is contributing to protecting the underlying Mg surface. This can be seen for pure Mg when investigated in various media, as the Nyquist plot clearly displays a second time constant (i.e. layer) in HBSS but not in a more simple NaCl solution (Fig. 2.6).

EIS may also be used to determine the protection offered by coatings placed on the Mg prior to corrosion, as well as determining when these layers break down [81]. This is especially crucial for the study of Mg in SBF, as CaP layers that form are one of the primary providers of protection to the Mg subsurface, and consequently, understanding their behaviour is necessary.

2.3.2.2 Drawbacks/Considerations for EIS

If the slopes of the individual reactions are known (from PDP), it is possible to apply the Stern–Geary equation to obtain an approximate i_{corr} from EIS data [83]. However, this equation relies on accurate determination (or alternatively, accurate assumption) of the Tafel slopes, which in turn requires correct analysis of polarisation data. For Mg in SBF, this may be challenging, as the anodic and cathodic slopes measured will alter if significant as corrosion occurs over time.

The impedance response is nominally analysed in terms of an equivalent circuit to quantify the resistance and capacitance. However, this can itself be inherently difficult, as often multiple equivalent circuits may fit the same data, resulting in considerably divergent calculated values (Fig. 2.7). Only the use of an equivalent circuit that accurately approximates the reactions and layers present at the surface will yield a meaningful interpretation of the data. Due to the diverse nature of the

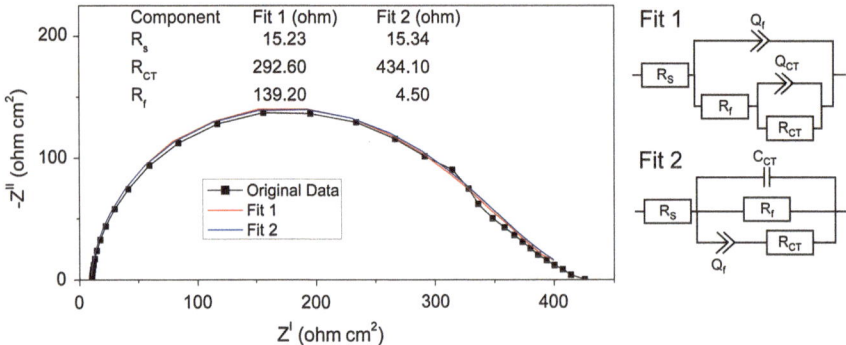

Fig. 2.7 EIS data for pure Mg fitted using two different equivalent circuits shown on the *right*. (R_s is solution resistance, R_f and Q_f are the resistance and constant phase element of a film on the sample surface, and R_{CT}, Q_{CT} and C_{CT} are the resistance, constant phase element and capacitance associated with charge transfer, respectively)

Mg corrosion layers, which depend heavily on the in vitro variables (e.g. media and buffer), correct choice of circuit requires some experience.

EIS cannot determine shifts instigated by different alloying elements or solutions. This makes it difficult to study the individual contributions from microstructural features (e.g. secondary phases). In addition, EIS does not directly yield a corrosion rate and is highly susceptible to any degradation, which can occur while the scan is running. Specifically, this degradation makes low-frequency measurements difficult, as the unaccelerated active reactions continuously occur, while the impedance/resistance is recorded (Fig. 2.8). This low-frequency behaviour is particularly strong for Mg in SBF, because of its relatively rapid rate of dissolution.

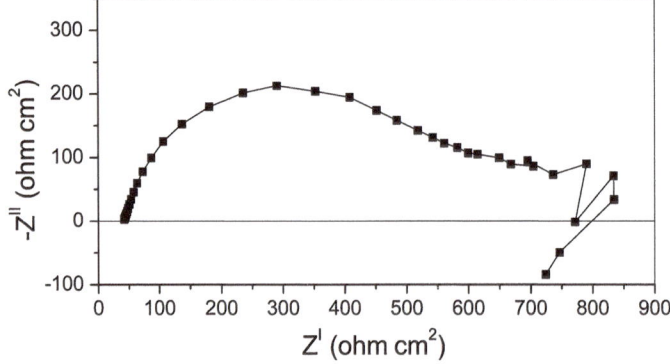

Fig. 2.8 Example of a Nyquist plot for pure Mg where active corrosion interferes with low-frequency behaviour as shown by the large scatter to the *right* of the plot

2.4 Conclusions

All the techniques discussed here are complementary to each other, with no experiment providing all of the information required to fully understand the corrosion behaviour of Mg alloys in an SBF. Investigators must be aware of the salient features of each technique.

Mass loss experiments provide a simple benchmark for determining the actual amount of cumulative corrosion that has occurred for Mg alloys in vitro. Although simple to set up, they cannot be accelerated, do not distinguish between general and non-uniform corrosion and reveal nothing about corrosion mechanisms.

Hydrogen evolution experiments allow a pseudo mass loss to be determined over time; however, the experiment is almost impossible to perform accurately. In addition, it may also not precisely identify the physical corrosion of many alloys (say, due to non-Faradaic dissolution, but in most cases, due to inefficient gas collection). Additionally, the majority of the literature has not reported the theoretical 1:1 ratio of hydrogen evolved to actual ML (Table 2.6).

Potentiodynamic polarisation allows quantification of the mechanisms of corrosion of Mg alloys. It provides kinetic information, allowing better understanding of why and how the corrosion is taking place. PDP, however, remains a single time-point test which destroys the sample surface, and for Mg in SBF, PDP results cannot accurately be converted to corrosion rates. PDP also does not allow in-depth analysis of the corrosion layers that form because of rapid dissolution of Mg in the bioelectrolytes used.

The primary benefit of electrochemical impedance spectroscopy is to elucidate the behaviour of corrosion layers that form. EIS can determine the time-dependant formation and dissolution of any layer at the Mg surface, providing quantitative analysis of the protection any layer provides. EIS, however, is readily affected by the continuing Mg dissolution at low frequencies, and the choice of electrical circuit to interpret the data can be controversial. Consequently, to be correctly used, EIS requires significant understanding of the corrosion processes that are taking place and how they are represented in collected data.

References

1. Prichard RW (1976) Animal models in human medicine. In: Animal models of thrombosis and hemorrhagic diseases, vol 76–982. US Department of Health, Education, and Welfare, Washington DC, pp 169–172
2. Bernard C (1957) An introduction to the study of experimental medicine (English Translation). Dover Publications, New York
3. National Research Council (1988) Use of laboratory animals in biomedical and behavioral research. National Academy Press, Washington DC
4. Matfield M (2002) Animal experimentation: the continuing debate. Nat Rev Drug Discov 1(2):149–152

5. Celarek A, Kraus T et al (2012) Phb, crystalline and amorphous magnesium alloys: promising candidates for bioresorbable osteosynthesis implants? Mater Sci Eng C 32(6):1503–1510
6. Chen S, Guan S et al (2012) In vivo degradation and bone response of a composite coating on Mg–Zn–Ca alloy prepared by microarc oxidation and electrochemical deposition. J Biomed Mater Res B Appl Biomater 100B(2):533–543
7. Duygulu O, Kaya RA et al (2007) Investigation on the potential of magnesium alloy AZ31 as a bone implant. Mater Sci Forum 546–549:421–424
8. Erdmann N, Angrisani N et al (2011) Biomechanical testing and degradation analysis of MgCa0.8 alloy screws: a comparative in vivo study in rabbits. Acta Biomater 7(3):1421–1428
9. Fischerauer SF, Kraus T et al (2013) In vivo degradation performance of micro-arc-oxidized magnesium implants: a micro-Ct study in rats. Acta Biomater 9(2):5411–5420
10. Heublein B, Rohde R et al (2003) Biocorrosion of magnesium alloys: a new principle in cardiovascular implant technology? Heart 89(6):651–656
11. Kräuse A, von der Höh N et al (2010) Degradation behaviour and mechanical properties of magnesium implants in rabbit tibiae. J Mater Sci 45(3):624–632
12. Kraus T, Fischerauer SF et al (2012) Magnesium alloys for temporary implants in osteosynthesis: in vivo studies of their degradation and interaction with bone. Acta Biomater 8(3):1230–1238
13. Krause C, Bormann D et al (2006) Mechanical properties of degradable magnesium implants in dependence of the implantation duration. In: Pekguleryuz M (ed) Conference of metallurgists: magnesium technology in the global age. Montreal, Quebec, pp 329–343
14. Li Z, Gu X et al (2008) The development of binary Mg-Ca alloys for use as biodegradable materials within bone. Biomaterials 29(10):1329–1344
15. Reifenrath J, Palm C et al (2005) Subchondral plate reconstruction by fast degrading magnesium scaffolds influence cartilage repair in osteochondral defects. In: Society OR (ed) 51st Annual Meeting of the Orthopaedic Research Society. Orthopaedic Research Society
16. Ren Y, Huang J et al (2007) Preliminary study of biodegradation of AZ31B magnesium alloy. Front Mater Sci China 1(4):401–404
17. Thomopoulos S, Zampiakis E et al (2009) Use of a magnesium-based bone adhesive for flexor tendon-to-bone healing. J Hand Surg 34(6):1066–1073
18. Von Der Höh N, Bormann D et al (2009) Influence of different surface machining treatments of magnesium-based resorbable implants on the degradation behavior in rabbits. Adv Eng Mater 11(5):B47–B54
19. Von Der Höh N, Krause A et al (2006) The influence of difference surface machining treatments of resorbable implants of different magnesium alloys: a primary study in rabbits. Biomaterialien 7(S1):122
20. Waksman R, Pakala R et al (2006) Safety and efficacy of bioabsorbable magnesium alloy stents in porcine coronary arteries. Cathet Cardiovasc Interventions 68(4):607–617
21. Waksman RON, Pakala R et al (2007) Efficacy and safety of absorbable metallic stents with adjunct intracoronary beta radiation in porcine coronary arteries. J Intervent Cardiol 20(5):367–372
22. Witte F, Abeln I et al (2008) Evaluation of the skin sensitizing potential of biodegradable magnesium alloys. J Biomed Mater Res: Part A 86A(4):1041–1047
23. Witte F, Fischer J et al (2006) Microtomography of magnesium implants in bone and their degradation. In: Progress in biomedical optics and imaging—proceedings of SPIE. International Society for Optical Engineering, Bellingham WA, WA 98227-0010, United States, p 631806
24. Witte F, Fischer J et al (2010) In vivo corrosion and corrosion protection of magnesium alloy LAE442. Acta Biomater 6(5):1792–1799
25. Witte F, Kaese V et al (2005) In vivo corrosion of four magnesium alloys and the associated bone response. Biomaterials 26(17):3557–3563
26. Witte F, Nellesen J et al (2006) In vitro and in vivo corrosion measurements of magnesium alloys. Biomaterials 27(7):1013–1018

27. Witte F, Reifenrath J et al (2006) cartilage repair on magnesium scaffolds used as a subchondral bone replacement. Materialwiss Werkstofftech 37(6):504–508
28. Witte F, Ulrich H et al (2007) Biodegradable magnesium scaffolds: part 2: peri-implant bone remodeling. J Biomed Mater Res: Part A 81A(3):757–765
29. Witte F, Ulrich H et al (2007) Biodegradable magnesium scaffolds: part 1: appropriate inflammatory response. J Biomed Mater Res: Part A:748–756
30. Wong HM, Yeung KWK et al (2010) A biodegradable polymer-based coating to control the performance of magnesium alloy orthopaedic implants. Biomaterials 31:2084–2096
31. Xu L, Pan F et al (2009) In vitro and in vivo evaluation of the surface bioactivity of a calcium phosphate coated magnesium alloy. Biomaterials 30(8):1512–1523
32. Xu L, Yu G et al (2007) In vivo corrosion behavior of Mg-Mn-Zn alloy for bone implant application. J Biomed Mater Res: Part A 83(3):703–711
33. Zhang EL, Xu LP et al (2009) In vivo evaluation of biodegradable magnesium alloy bone implant in the first 6 months implantation. J Biomed Mater Res: Part A 90A(3):882–893
34. Zhang S, Zhang X et al (2010) Research of Mg-Zn Alloy as degradable biomaterial. Acta Biomater 6(2):626–640
35. Bosiers M (2009) AMS insight—absorbable metal stent implantation for treatment of below-the-knee critical limb ischemia: 6 month analysis. Cardiovasc Intervent Radiol 32(3):424–435
36. Erbel R, Di Mario C et al (2007) Temporary scaffolding of coronary arteries with bioabsorbable magnesium stents: a prospective, non-randomised multicentre trial. The Lancet 369(9576):1869–1875
37. McMahon CJ, Oslizlok P et al (2007) Early restenosis following biodegradable stent implantation in an aortopulmonary collateral of a patient with pulmonary atresia and hypoplastic pulmonary arteries. Catheterization and Cardiovascular Interventions 69(5):735–738
38. Peeters P, Bosiers M et al (2005) Preliminary results after application of absorbable metal stents in patients with critical limb ischemia. J Endovasc Ther 12(1):1–5
39. Schranz D, Zartner P et al (2006) Bioabsorbable metal stents for percutaneous treatment of critical recoarctation of the aorta in a newborn. Cathet Cardiovasc Interv 67(5):671–673
40. Zartner P, Cesnjevar R et al (2005) First successful implantation of a biodegradable metal stent into the left pulmonary artery of a preterm baby. Cathet Cardiovasc Interv 66:590–594
41. Ren Y, Wang H et al (2007) Study of biodegradation of pure magnesium. Key Eng Mater 342–343:601–604
42. Wang RR, Li Y (1998) In vitro evaluation of biocompatibility of experimental titanium alloys for dental restorations. J Prosthet Dentist 80(4):495–500
43. Black J (2006) Biological performance of materials: fundamentals of biocompatibility, 4th edn. Marcel Dekker, New York
44. ASTM International (2004) ASTM Standard G31-72, "Standard Practice for Laboratory Immersion Corrosion Testing of Metals". ASTM International, West Conshohocken
45. Witte F, Feyerabend F et al (2007) Biodegradable magnesium-hydroxyapatite metal matrix composites. Biomaterials 28(13):2163–2174
46. Eliezer A, Witte F (2010) Corrosion behaviour of magnesium alloys in biomedical environments. Adv Mater Res 95:17–20
47. Auer J, Goodship A et al (2007) Refining animal models in fracture research: seeking consensus in optimising both animal welfare and scientific validity for appropriate biomedical use. BMC Musculoskelet Disord 8(1):72
48. Gill TJ, McCulloch PC et al (2005) Chondral defect repair after the microfracture procedure: a nonhuman primate model. Am J Sports Med 33(5):680–685
49. Makar GL, K J (1993) Corrosion of magnesium. Int Mater Rev 38(3):138–153
50. Shi Z, Atrens A (2011) An innovative specimen configuration for the study of Mg corrosion. Corros Sci 53(1):226–246
51. Winston R, Herbert R et al (2008) Thermodynamics: Pourbaix diagrams. In: Corrosion and corrosion control (4th edn). pp 43–51

52. Lorenz C, Brunner JG et al (2009) Effect of surface pre-treatments on biocompatibility of magnesium. Acta Biomater 5(7):2783–2789
53. Ng WF, Chiu KY et al (2010) Effect of pH on the in vitro corrosion rate of magnesium degradable implant material. Mater Sci Eng C 30(6):898–903
54. Yang L, Zhang E (2009) Biocorrosion behavior of magnesium alloy in different simulated fluids for biomedical application. Mater Sci Eng C 29(5):1691–1696
55. Yamamoto A, Hiromoto S (2009) Effect of inorganic salts, amino acids and proteins on the degradation of pure magnesium in vitro. Mater Sci Eng C 29(5):1559–1568
56. Li L, Gao J et al (2004) Evaluation of cyto-toxicity and corrosion behavior of alkali-heat-treated magnesium in simulated body fluid. Surf Coat Technol 185(1):92–98
57. Wang H, Shi ZM et al (2008) Magnesium and magnesium alloys as degradable metallic biomaterials. Adv Mater Res 32:207–210
58. Vojtěch D, Čížová H et al (2006) Investigation of magnesium-based alloys for biomedical applications. Kovove Mater 44:211–223
59. Hort N, Huang Y et al (2010) Magnesium alloys as implant materials: principles of property design for Mg-Re alloys. Acta Biomater 6:1714–1725
60. Song G, Atrens A et al (2001) An hydrogen evolution method for the estimation of the corrosion rate of magnesium alloys. In: Hyrn JN (ed) Magnesium technology 2001 symposium. Minerals, Metals and Materials Society, New Orleans, pp 255–262
61. Xin Y, Liu C et al (2007) Corrosion behavior of biomedical AZ91 magnesium alloy in simulated body fluids. J Mater Res 22(7):2004–2011
62. Zhang CY, Zeng RC et al (2010) Comparison of calcium phosphate coatings on Mg-Al and Mg-Ca alloys and their corrosion behavior in Hank's solution. Surf Coat Technol 204(21–22):3636–3640
63. Wilhelm E, Battino R et al (1977) Low-pressure solubility of gases in liquid water. Chem Rev 77(2):219–262
64. Baranenko VI, Kirov VS (1989) Solubility of hydrogen in water in a broad temperature and pressure range. At Energ 66(1):30–34
65. Crozier TE, Yamamoto S (1974) Solubility of hydrogen in water, sea water, and sodium chloride solutions. J Chem Eng Data 19(3):242–244
66. Barton RS (1960) The permeability of some plastic materials to H_2, He, N_2, O_2, and Ar. atomic energy research establishment report
67. Piskarev I, Ushkanov V et al (2010) Establishment of the redox potential of water saturated with hydrogen. Biophysics 55(1):13–17
68. Denkena B, Lucas A (2007) Biocompatible magnesium alloys as absorbable implant materials: adjusted surface and subsurface properties by machining processes. CIRP Ann Manufact Technol 56(1):113–116
69. Song G (2005) Recent progress in corrosion and protection of magnesium alloys. Adv Eng Mater 7(7):563–586
70. Shi Z, Liu M et al (2010) Measurement of the corrosion rate of magnesium alloys using Tafel extrapolation. Corros Sci 52(2):579–588
71. Kirkland NT, Williams G et al (2012) Observations of the Galvanostatic dissolution of pure magnesium. Corros Sci 65:5–9
72. Birbilis N, Easton MA et al (2009) On the corrosion of binary magnesium-rare Earth alloys. Corros Sci 51(3):683–689
73. Kirkland NT, Staiger MP et al (2011) Performance-driven design of biocompatible Mg-alloys. JOM 63(6):28–34
74. Kirkland NT, Birbilis N et al (2010) In-vitro dissolution of magnesium–calcium binary alloys: clarifying the unique role of calcium additions in bioresorbable magnesium implant alloys. J Biomed Mater Res B Appl Biomater 95B(1):91–100
75. ASTM International (2004) ASTM Standard G102-89, standard practice for calculation of corrosion rates and related information from electrochemical measurement. ASTM International, West Conshohocken

76. Tait WS (1994) An introduction to electrochemical corrosion testing for practicing engineers and scientists. PairODocs Publications, Racine
77. Wang J (2002) Analytical electrochemistry, 2nd edn. Wiley, New York
78. Zhang XL, Jiang ZH et al (2009) Effects of scan rate on the potentiodynamic polarization curve obtained to determine the Tafel slopes and corrosion current density. Corros Sci 51(3):581–587
79. Gosser DK (1993) Cyclic voltammetry: simulation and analysis of reaction mechanisms. Wiley, New York
80. MacDonald DD, McKubre MCH (1987) Impedance measurement techniques. In: Macdonald JR (ed) Impedance spectroscopy: emphasizing solid materials and systems. Wiley-Interscience, New York, p 133
81. Ghali E (2010) Conventional and electrochemical methods of investigation. In: Corrosion resistance of aluminium and magnesium alloys: understanding, performance and testing. Wiley, Hoboken
82. Lasia A (1999) Electrochemical impedance spectroscopy and its applications. In: Conway BE, Bockris J, White RE (eds) Modern aspects of electrochemistry, vol 32. New York, pp 143–248
83. Tern M, Geary AL (1957) Electrochemical polarization: a theoretical analysis of the shape of polarization curves. J Electrochem Soc 104(1):56–63

Chapter 3
Influence of Environmental Variables on In Vitro Performance

Abstract To date, many in vitro Mg biodegradation tests are not being carried out in a reproducible or clear manner. In many cases, values of important variables (such as pH) are not reported, or are not correctly controlled throughout the experiments; in other cases, values are being used which do not mimic physiological conditions. Typical issues include the use of unphysiological temperatures or alkalised media, as well as set-ups with either unadjusted or uncontrolled pH values, far outside the body's natural range. In many studies, it is difficult to determine what was and was not controlled—this also makes it difficult to compare studies. The influences and impacts of experimental parameters such as pH, solution composition and temperature are discussed in this chapter. It is shown that it is very difficult to relate results from studies performed under non-physiological conditions to in vivo performance.

Keywords Variable · Temperature · Simulated body fluid · pH · Buffer · In vitro · In vivo · Protein · Amino acid

3.1 Introduction

An informal survey conducted by the authors of in vitro studies reported in the literature to date found that over 75 % of these studies failed to mention and/or properly control single or multiple variables that could significantly affect the results. In some studies, the values of important variables were simply not reported, making it impossible to repeat the experiments or to determine what was controlled and within what range [1, 2]. Errors in other studies included the use of extremely basic media as "simulated body fluids" [3, 4], employing unphysiological temperatures [5, 6] and not monitoring or controlling pH values to within the body's natural range [7, 8].

Before defining recommended standard values and ranges for in vitro variables, we need to first understand how each variable affects the results obtained from

N. T. Kirkland and N. Birbilis, *Magnesium Biomaterials*, SpringerBriefs in Materials, DOI: 10.1007/978-3-319-02123-2_3, © The Author(s) 2014

different experiments and, therefore, how important control of each variable is. In this chapter, we examine the most important variables, which are common to all in vitro biocorrosion experiments, including temperature, pH, buffering system, choice of simulated body fluid and sample preparation technique. Some of these variables, such as temperature, are simple to control, and an ideal value can be provided. However for others, including choice of medium and buffering system, greater understanding of the specific effects of the variable on degradation mechanisms is required to make an appropriate selection during experimental design.

3.2 Temperature Effects

Temperature can affect the dissolution and corrosion of metals in several ways. In the case of Mg in simulated body fluid, corrosion proceeds via the metal oxidation process, with no significant passivation (neglecting any surface phosphates, or coatings). As such, the corrosion rate will increase exponentially with increasing temperature, following an Arrhenius-type relationship [9]. In other, passive metal systems, temperature also affects the extent of localised corrosion by influencing the pitting potential (viz. critical pitting temperature). Passive metals used in biomedical applications include stainless steels [10–12] and titanium alloys [13]. Further, for alloyed materials, the individual electrochemical kinetics of constituent phases also vary quite dramatically with temperature, as shown rather elegantly for Al alloys [14]. These important second-order effects are not always relevant for Mg (which dissolves freely at neutral pH), although they may occur in alloyed materials. Temperature also affects the Gibbs free energy of adsorption for proteins and thus affects the affinity and number of proteins that adsorb to a surface [11]. In addition, temperature will play a significant role in the survival of cells during in vitro testing, an effect that has been appreciated for over a century [15].

The rate of the dissolution (oxidation) reaction will increase with increasing temperature, leading to an approximate doubling of the corrosion rate for each 10 °C at temperatures around ambient. Variations in temperature for tests reported to date make drawing extended conclusions, or making definite correlations between like studies, very difficult. Examples of in vitro tests where temperature was not held at the physiological temperature (T_{phys}) are given in Appendix C. Of the literature cited in this book, approximately 40 % of studies have not carried out their in vitro corrosion testing (such as mass loss, electrochemistry) at T_{phys}.

This discrepancy would not be such a problem if there were a simple pattern to the differences between results collected at room temperature and those obtained at T_{phys}. Unfortunately, making this comparison from the above-cited works is almost impossible. This is because corrosion is evaluated using a variety of metrics such as i_{chor}, polarisation resistance (R_p) and mass loss (often as a percentage change). To make comparisons, one would need to somehow normalise the corrosion rates for specimen geometry and exposure time—but often specimen geometry (and

Fig. 3.1 Mass loss rates of alloys tested at 20 ± 0.5 °C and 37 ± 0.5 °C after 7 days exposure

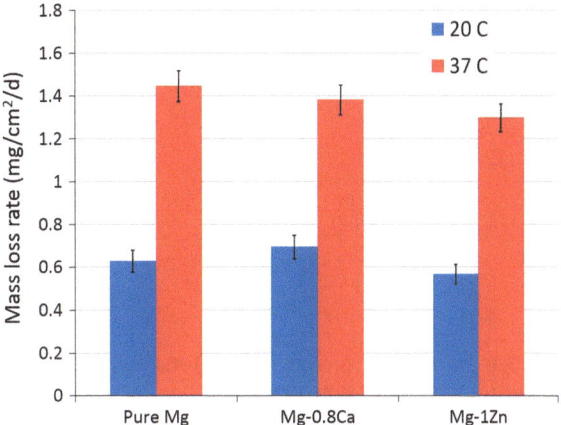

associated parameters such as mass of original specimens and electrode areas) is not actually reported. Therefore, we have instead chosen to make some key comparisons that illustrate important effects and differences.

An example of the effect of temperature on the mass loss determined following in vitro exposure is shown in Fig. 3.1, which shows that the mass loss is markedly different depending on the temperature of the Hanks' balanced salt solution (HBSS). The temperature change seems to at least double the mass loss, implying that studies that are not executed at T_{phys} are incorrect by at least a factor of two. It is also important to note that the amount of change in mass loss with temperature is different for each alloy, meaning that assertions from data not at T_{phys} will not necessarily translate correctly to physiological conditions. At 20 °C, the pure Mg has a lower mass loss than that of Mg-0.8Ca; however, at T_{phys}, the opposite is true.

The classical electrochemical influence of temperature on dissolution is elegantly captured in Fig. 3.2. The rate of the anodic reaction increases with increasing temperature. The electrochemical impact is confined (nominally exclusively) to the anodic reaction, since the cathodic reaction is an electron-transfer reaction. There are a number of important points to be made from inspection of Fig. 3.2, including the following:

1. The change in the anodic reaction is very large.
2. The rate of change in kinetics is alloy dependent. In the alloy-rich examples, Mg-10Ca (wt. %) and Mg-10Zn (wt. %), the increase in anodic kinetics is greater for the Mg-10Ca (wt. %) specimen. These alloys have significant fractions of second-phase particles [16], and this difference in behaviour arises because the activation of the Mg_2Ca intermetallic is more rapid than that of $MgZn_2$. As such, results at non-T_{phys} are irrelevant when considering applications within the body.
3. The increased rates of anodic reactions at T_{phys} are accompanied by a lowering of the corrosion potential (E_{corr}), such that the alloy potentials reported at 20 °C

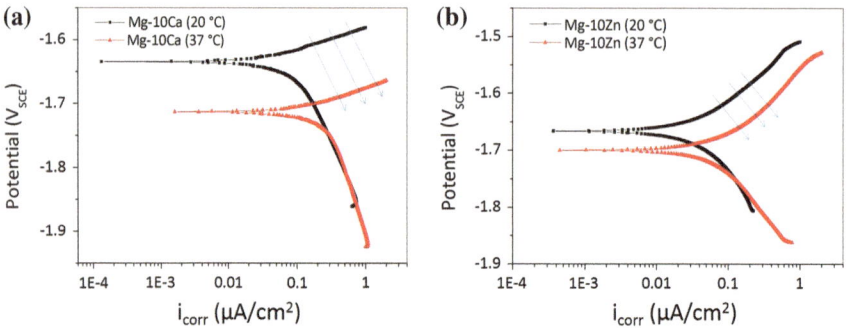

Fig. 3.2 Polarisation curves for **a** Mg-10Ca (wt.%) and **b** Mg-10Zn (wt.%) at 20 and 37 °C. Anodic shift is indicated by *arrows*

are necessarily (and often significantly) more noble. This has a further ramification, in that the rate of hydrogen evolution can be influenced by the alloy potential as the potentials tend towards $-1V_{SHE}$. This is mostly relevant to heavily alloyed Mg (which is ennobled) or Mg-based metallic glasses (which are very heavily alloyed). It is important to remember that results (particularly those relating to hydrogen evolution) at 20 °C will not translate to performance at T_{phys}.

Temperature may also affect the accuracy of the experiment itself, as is the case for hydrogen evolution experiments, where the solubility of hydrogen in media varies significantly between room and T_{phys} (See Sect. 3.2).

Selected results for a number of alloys tested in either HBSS or minimum essential medium containing 10 % foetal bovine serum (MEM$_{FBS}$) are presented in Fig. 3.3. All alloys displayed increased i_{corr} at the higher temperature; the changes ranged from 64 to 840 % in HBSS and 14–437 % in MEM$_{FBS}$. The AZ alloys, which displayed the lowest corrosion rates at room temperature, showed much

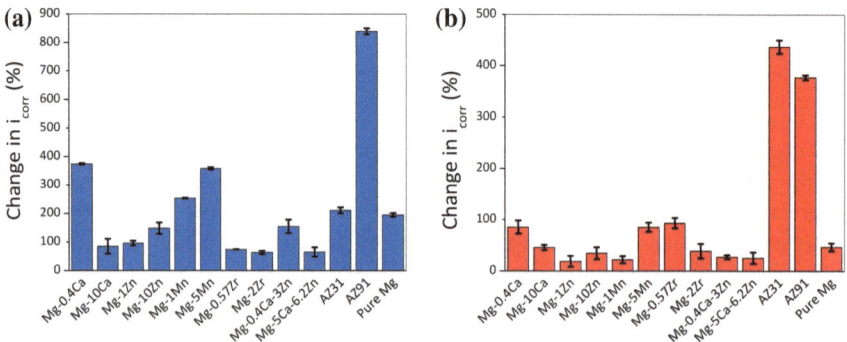

Fig. 3.3 Δi_{corr} upon increase in temperature from 20 to 37 °C for various alloys in **a** HBSS and **b** MEM$_{FBS}$

larger increases in corrosion rate with temperature (such as an eightfold increase for AZ91). Comparison with the relatively small increase in i_{corr} of 14 % for Mg-1Zn highlights the variation between alloy measurements at different temperatures and in different solutions, further demonstrating that conclusions from room temperature experiments are likely to be inaccurate at T_{phys}.

A possible reason for the acceleration in corrosion (above and beyond the factor of two expected for the increased anodic reaction kinetics) is an increased rate of diffusion of hydrogen ions (H^+) at the higher temperature, which would then also increase the amount of H_2 gas that forms at the surface. It has been proposed that the process of Mg corrosion is primarily controlled by the transport rate of hydrogen on the surface (cathodically limited, in mildly acidic solutions) and thus the increase in diffusion of hydrogen is the main reason for increasing corrosion [17]. However, if this were the case, the corrosion should be cathodically controlled; but many alloys (such as those shown in Fig. 3.2) do not display cathodically controlled corrosion.

3.3 Buffering Systems and Effect of Solution pH on Mg Biocorrosion

pH, originally defined by Danish biochemist Søren Peter Lauritz Sørensen in 1909, is a measure of the "power of hydrogen" in the system. It is the negative logarithm of the H^+ concentration, as shown in

$$pH = -\log_{10}(\alpha_{H^+}) \tag{3.1}$$

where α_{H^+} is the activity (related to concentration) of H^+.

3.3.1 pH Role in General Mg Corrosion

The crucial role that pH plays in the corrosion of Mg and its alloys has been well documented [18, 19]. When placed in a medium, Mg will form a magnesium hydroxide film that can provide some form of protection over a wide range of pH values. However, the formation and effectiveness of this film is readily hindered in even mildly aggressive media or in the presence of impurities. Consequently, Mg will often dissolve relatively rapidly, unless the surface pH remains higher than 11 (the minimum required value for $Mg(OH)_2$ to be stable). At lower pH values, a less thick and more fragile surface layer will form, providing limited protection. Thus, the pH of the solution surrounding an Mg sample is crucial to the formation of the $Mg(OH)_2$ layer. At values close to those of the body, 7.4–7.6, this layer is neither stable nor complete and will continue to dissolve in any media similar to body fluids.

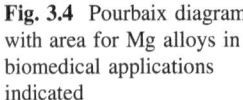

Fig. 3.4 Pourbaix diagram with area for Mg alloys in biomedical applications indicated

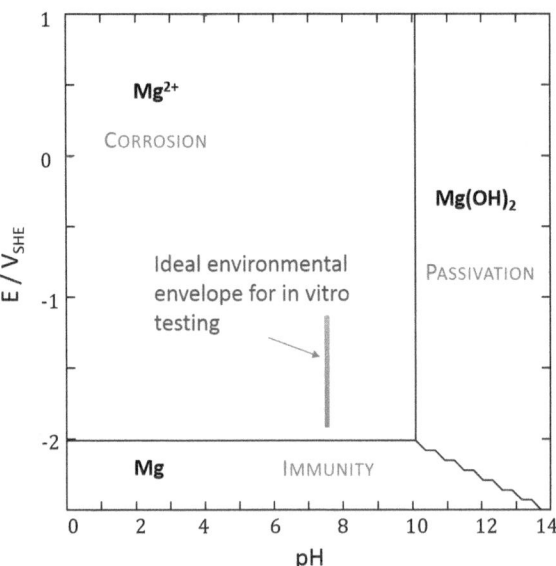

Mg and its alloys for biological applications occupy only a small part of the Pourbaix diagram, because of the range of physiological pH values allowed and due to the potential being limited to the corrosion potentials of the various alloys (Fig. 3.4).

3.3.2 pH Role in Biocorrosion

The role of pH is even more complicated and important in experiments intended to simulate biocorrosion in vitro. A higher pH has been shown to result in an increased and thicker formation of calcium phosphate (CaP) compounds (such as hydroxyapatite) [20, 21]. This is important for a number of reasons. First of all, these layers can provide additional protection to the underlying surface, reducing corrosion. Secondly, the components of these layers are the precursors to bone grown in vivo, and thus studying their formation in vitro may be crucial to understanding the performance of potential orthopaedic implants.

pH is also known to play a significant role in the adsorption of proteins [22]. Proteins will be adsorbing to the layers that form on the Mg surface, in this case CaP, and pH has been shown to significantly affect the numbers and affinity of proteins which adsorb to CaP [23]. Proteins will influence the attachment of cells to the biomaterial surface, affecting not only corrosion but the biocompatibility and ultimately the success of the implant. It should be noted that CaP layers, often observed in more realistic SBFs that contain the necessary Ca^{2+} and PO_4^{3-}, are not found in NaCl solutions [5, 24, 25], highlighting the importance of medium choice, which is discussed further in Sect. 3.4.

Of the literature reviewed by the authors, less than 40 % of studies both adjusted the pH to physiological levels and maintained a suitable range throughout longer-term biocorrosion experiments. The vast majority either did not adjust the pH, did not maintain it, or failed to report pH control in their work. This highlights a significant failing for the community as a whole and severely limits the use of the data that have been reported.

It is clear that investigation of the effects that pH can have on the degradation rates and mechanisms of Mg and its alloys is needed. Consequently, the authors sought to clarify how pH can affect not only the corrosion rate of Mg but also the morphology of corrosion layers that form on its surface. When considering pH, it is also necessary to think about the roles that the buffering agents can play.

It is important to consider that some have suggested that shortly after implantation of an implant (in a bone fracture site), the pH of the surround tissues can drop to 5.5 [26]. However, pH changes because of surgery are ordinarily only temporary, unless significant tissue damage occurs during surgery or the implant is not well received [27]. Further, it has been found that, after an orthopaedic operation, pH in the vicinity of the implant varied only ± 0.16 around the baseline value of 7.4 during the 30 post-operative days [28], with variation attributed to tissue repair during bone healing.

3.3.3 Buffers and pH Control

Buffer solutions essentially operate as neutralising agents, containing both a weak acid and its conjugate base. pH is kept constant when more acid is added to the solution, because the H^+ ions can combine with the conjugate base. Similarly, OH^- ions will accept protons from the weak acid (forming H_2O or another compound that has little effect on the pH of the system as a whole) to maintain the pH. Each buffer has a buffering capacity, which is the maximum amount of H^+ or OH^- that the buffer can accept. In more complex media, other components such as amino acids (AA) also act as buffers [29].

The human body relies on a buffering system that uses bicarbonate ions, carbon dioxide (CO_2) and carbonic acid. These components work together to neutralise hydroxide and H^+ which might otherwise alter the pH [30]. This control is vital to the normal function of proteins and cells, and the body is normally effective at maintaining a suitable pH range by triggering the release of CO_2 via respiration when pH levels drop [31].

There are a number of buffer systems available which are specifically designed for use in in vitro experiments. Of these, the most realistic (i.e. similar to the body's natural mechanism) is a combination of sodium bicarbonate ($NaHCO_3$) and a controlled partial CO_2 atmosphere. In most in vitro experiments, 2.2 g/L of $NaHCO_3$ is added to an SBF that is kept in a 5–10 % CO_2 environment. This is effective at keeping the pH balanced and is a frequently used method in cell culture, where control of the pH value is crucial to avoid cell death. The primary benefit of the $NaHCO_3$ and CO_2 (SB_{CO_2}) system is its similarity to the in vivo buffering mechanisms. It is also relatively cheap and can provide chemical benefits (not found with other buffers) to the cells [32]. However, normally an incubator is required to provide the correct environment, and special vented flasks must be used to allow the CO_2 to regulate the pH.

The second principal type of buffering systems is based on a zwitterion, a highly water-soluble molecule that has both positive and negative charges and naturally contains both an acid and base. The most commonly used of these buffers is 4-(2-hydroxyethyl)-1-piperazineethanesulfonic acid, or HEPES. First described in the 1960s for maintaining pH during cell cultures, HEPES has since become widely popular [33]. The primary benefit of HEPES is the lack of requirement for a CO_2 environment, allowing studies to be performed in a variety of situations. While high concentrations (>50 mmol/L) of HEPES have been shown to be toxic to certain cell groups [32, 34], other studies have shown that HEPES strongly supports cell growth, with results comparable to that in SB_{CO_2} in some situations [35].

Other chemical buffers include tris(hydroxymethyl)aminomethane (TRIS), borate buffer and citric acid, although these have been used in less than 10 % of the Mg biocorrosion literature. Borate and citric acid buffers are used for pH control outside the physiological range, and as such warrant little further investigation.

Our examination of the literature found that over 70 % of studies either did not use or did not mention the use of a buffer. As with adjustment and control of the pH, this is a significant concern, especially in longer-term studies. Even if the test media are replaced at controlled intervals, it is still quite likely that the pH will rise significantly in between these changes, creating unphysiological conditions. Additionally, if there is no flow, the pH near the surface may be different to that of the bulk solution, and this effect will be magnified when no buffer is used. Even for short-term experiments, where it might be assumed that a buffer is not necessary because of the time-frame of the test, the buffer will act to counterbalance changes in pH that occur very soon after the sample is submerged, and thus, use of a buffer is still beneficial in short-term tests.

We will now examine the differences in biocorrosion results obtained using the most common HEPES and the most realistic (SB_{CO_2}) buffering systems. Our aim is to determine the role of buffering system in corrosion of Mg.

3.3.4 EIS of Pure Mg with Varying pH

Resistance values obtained from EIS experiments on Mg in HBSS adjusted to different pH values are presented in Fig. 3.5. Intermediate-pH levels of between 7.2 and 7.6 resulted in resistance values within 20 % of each other, while the more acidic pH of 7.0 resulted in a significantly smaller resistance after 24 h. The most dramatic difference occurred when the pH was increased to 7.8. At this pH, the initial resistance was significantly greater, and the resistance continued to increase after the first three hours at a rate approximately double that of the three intermediate-pH samples. The final resistance value ($\sim 1,800\ \Omega/cm^2$) was about 70 % higher than that of the intermediate-pH samples, indicating that, at this pH, corrosion is expected to be much slower.

To better understand the reasons for these differences, the contributions of the individual resistance components of the Mg surface and any associated film were investigated (Fig. 3.6). This resistance stabilises somewhat after 14 h of

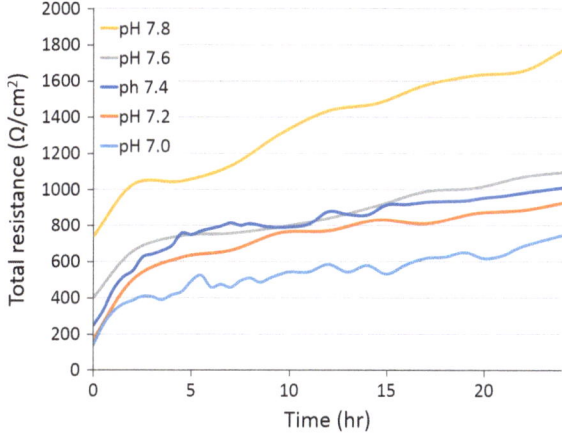

Fig. 3.5 R_{tot} of pure Mg immersed in HBSS with pH values between 7 and 7.8 over 24 h

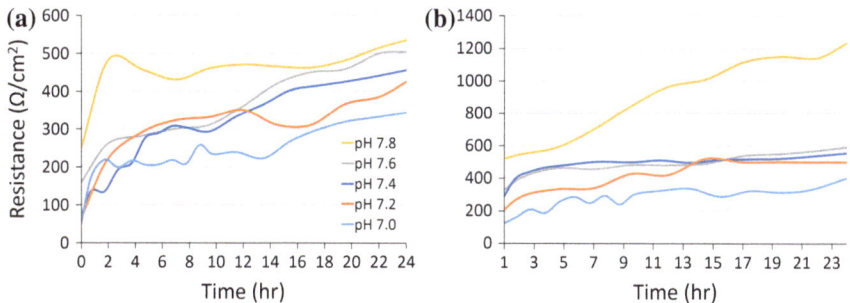

Fig. 3.6 **a** R_{CT} and **b** R_f obtained for different pH solutions (7.0–7.8) over 24 h

immersion, with all R_{CT} values within 200 Ω/cm^2 of each other, indicating that pH has a relatively weak influence on this parameter.

However, the film resistance (R_f), which represents the resistance of any significant layer that forms between the Mg and the solution, is more readily affected by pH. The three intermediate-pH solutions quickly stabilised to a fairly constant R_f value of 500–600 Ω/cm^2 (Fig. 3.6b). This indicates that the surface layer, once initially formed, probably does not become significantly thicker or denser and thus provide further coverage of the bare Mg. A pH value of 7.0 resulted in a marginally lower R_f of $\sim$400 Ω/cm^2. However, a pH of 7.8 resulted in a rapidly increasing R_f value after just five hours of immersion. Comparing the R_f and R_{CT} values implies that an increasingly protective coating is the main reason for the steady rise in total resistance at a pH of 7.8.

One reason suggested for this behaviour is that a higher concentration of OH$^-$ groups near the surface at higher pH values will make it more difficult for Cl$^-$ ions to adsorb onto the surface, by repelling the Cl$^-$ from the solution/surface interface [19]. This means that the effect of Cl$^-$ on corrosion rate is reduced in an OH$^-$ rich solution (i.e. Cl$^-$:OH$^-$ ratio); however, this has not been analysed in more detail.

3.3.5 PDP of Mg at Different pH Values

A similar trend was observed from potentiodynamic polarisation results, with approximately 75 % more corrosion occurring at a pH of 7.0 than in the 7.2–7.6 range (Fig. 3.7). From pH 7.2 to 7.6, the i_{corr} decreased by only 6 %, within one standard deviation. At pH 7.8, the average i_{corr} was 49 μA/cm^2—40 % less than that observed at pH 7.6.

To better understand the reasons for these differences, scanning electron microscope (SEM) images were taken of the Mg surfaces after immersion in

Fig. 3.7 i_{corr} of pure Mg in HBSS as a function of pH

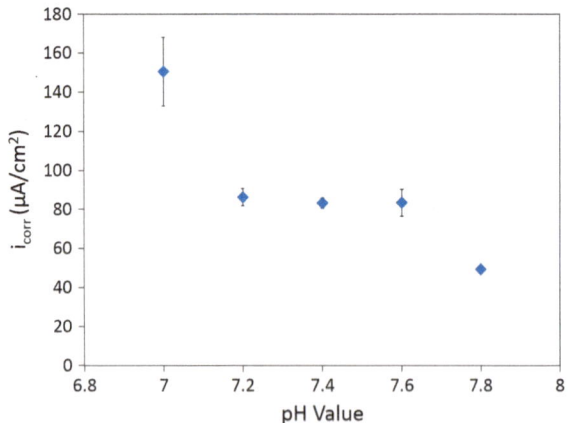

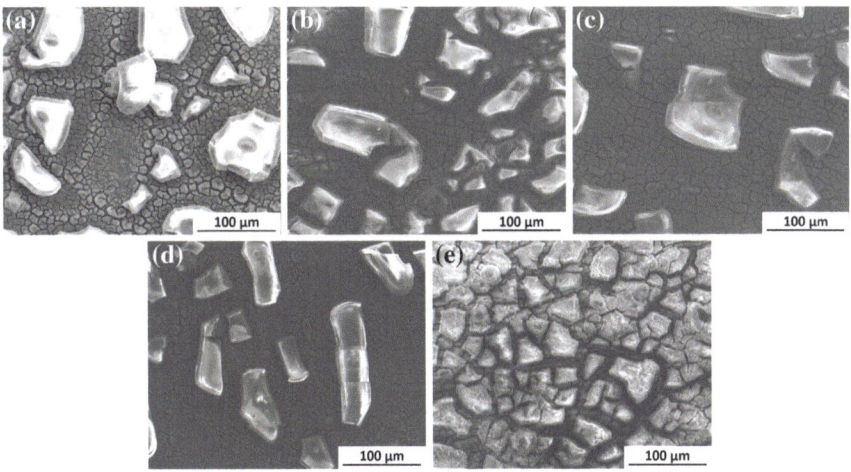

Fig. 3.8 Scanning electron micrographs of the surfaces of pure Mg samples after 24-h immersion in HBSS at pH **a** 7.0, **b** 7.2, **c** 7.4, **d** 7.6, and **e** 7.8

solutions of different pH (Fig. 3.8). Energy-dispersive X-ray spectroscopy (EDS) was also used to determine the composition of the surface (Fig. 3.9). For all samples, a dry, cracked-earth or clay-like layer was observed, composed of Mg and O. This is the $Mg(OH)_2$ layer that forms on Mg when placed in aqueous solutions (and acquires a clay-like appearance when dried before SEM analysis). Above this layer, samples also displayed a secondary layer comprised primarily of Mg, with increasing amounts of Ca, phosphorus (P) and O, as the pH increased. This indicates the formation of CaPs, which are known to preferentially nucleate on $Mg(OH)_2$ [36].

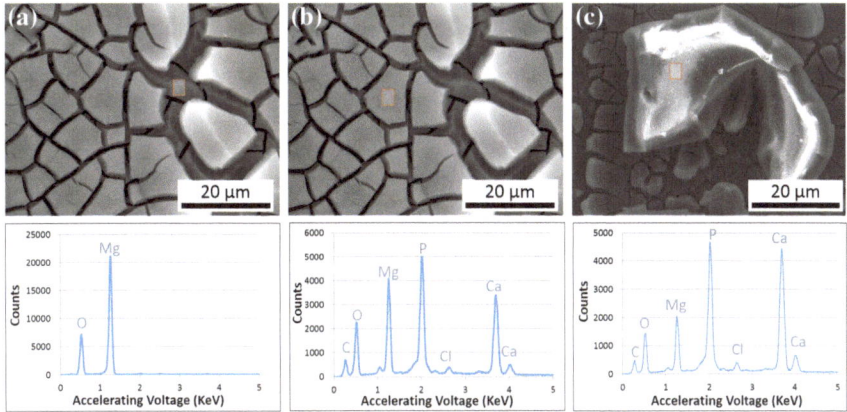

Fig. 3.9 Scanning electron micrographs and EDS analysis of typical surfaces on pure Mg at pH 7.0–7.6 showing; **a** $Mg(OH)_2$ underlayer, **b** $Mg(OH)_2$/CaP mixed layer, and **c** CaP flake

However, the elemental analysis of the layers in the pH 7–7.6 solutions implied that the layers were a mixture of Mg and CaP, probably because of the presence of either a thinner CaP layer and/or a combination of CaP and $Mg(OH)_2$. Large "flakes" comprised entirely of Ca, P and O were found on all surfaces except the pH 7.8 sample. These flakes were crystalline in appearance, from 40 to 100 μm across, and generally showed poor coverage of the surface below. In contrast, the surface of the pH 7.8 sample was completely covered in a CaP coating, with Mg only detected in the cracks within the coating.

This difference in the structure and makeup of the surface films between the different pH values is the main cause of the changes in corrosion rate. Only the pH 7.8 samples displayed a thick, CaP-based coating, shown by EIS to provide the bulk of the corrosion protection (Fig. 3.6). To investigate this effect further, a chemical equilibrium software package was used to determine the idealised formation conditions for CaP across a wide range of pH values (Medusa [37]). In the programme, the temperature was set to T_{phy} and the concentrations of Ca^{2+} and PO_4^{3-} were set at those found in HBSS. This allowed the creation of a diagram showing the fraction of CaP formation versus pH (Fig. 3.10).

According to the fraction equilibrium diagram, the CaP formed should be primarily one of the two forms: calcium-deficient apatite (CDA, Ca^{2+}/PO_4^{3-} ratio 1.33–1.66) or hydroxyapatite (HA, Ca^{2+}/PO_4^{3-} ratio 1.67) [21]. Note that both of these forms have Ca^{2+}/PO_4^{3-} ratios close to that of HBSS (1.625).

From Fig. 3.10, it is clear that the physiological pH range is crucial in the formation of CDA/HA. Even small changes in pH in this range can result in large shifts in the amount of CaP formed. However, the equilibrium equations used for this graph assume perfect mixing and an infinite time, which is probably why the actual differences in corrosion values (and assumed formation of CaP layers on the surface) do not precisely correlate with the fraction of CaP that ideally forms at each pH value. Additionally, other CaP compounds ($CaHPO_4$, $Ca(H_2PO_4)$, etc.) may have also formed in the pH experiments, because the solutions are not an entirely ideal environment. Moreover, it has been found that Mg itself may substitute for Ca in certain situations, creating carbonated magnesium-CaPs [38].

Fig. 3.10 CaP formation according to ideal chemical equilibrium reactions over a pH range of 4–10 in HBSS-equivalent solution

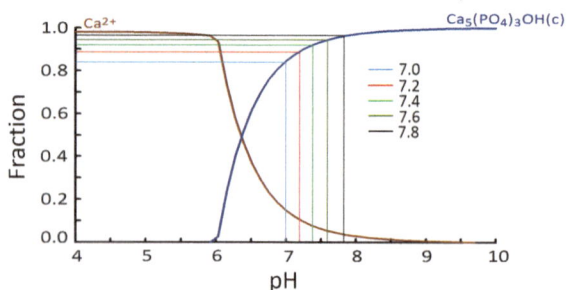

3.3.6 Effect of Buffer in Immersion Tests

To investigate the effect that a buffer choice can have on the biocorrosion of Mg, tests were carried out in the most common zwitterionic buffer HEPES, the most realistic buffer (SB_{CO_2}), and a combination of both based on a commonly available dual-buffer system known as "Dutch Modification" [39]. Samples were immersed in a variety of SBFs, including HBSS, Earle's balanced salt solution (EBSS), minimum essential medium (MEM) and an MEM containing 40 g/L of bovine serum albumin (MEM_{BSA}).

Mass loss data demonstrated a significant increase in the corrosion of the pure Mg in every solution when the buffering system was changed from SB_{CO_2} to HEPES, with 100–350 % more corrosion occurring (Fig. 3.11). This was further increased when both buffers were used in combination, resulting in a 300–620 % increase in corrosion over the SB_{CO_2}-buffered solutions. Thus, the choice of buffer plays a vital role in the corrosion of Mg; in this case, the choice of buffer has a far greater effect than the choice of medium itself. Excluding the BSA-containing solution, all the media buffered with SB_{CO_2} displayed a virtually indistinguishable mass loss of 1.6 ± 0.2 %. For the HEPES-buffered media, this range was greater at 7 ± 1 %. Overall, there was a mean increase in corrosion of ~ 350 % for samples in HEPES-buffered solutions when compared to those buffered with SB_{CO_2}.

To better understand the observed differences in corrosion rate, SEM images were taken of the sample surfaces after three hours immersion (Fig. 3.12). Comparable surface layers were found, displaying the characteristic cracked-earth-like morphology commonly reported in the literature [5, 40, 41]. EDS analysis of the HEPES-buffered samples found the surface entirely composed of Mg and O, indicative of $Mg(OH)_2$ (as found previously for a variety of pH values). The surfaces of samples that had been in the solution buffered with SB_{CO_2} displayed a much denser layer, containing ~ 10 at. % Ca and P. A sublayer was also visible between the cracks of this film, composed almost entirely of Mg and O. This data indicate that even at the early stages of immersion, a CaP layer is rapidly formed in SB_{CO_2}-buffered media; this layer likely provides significant protection from corrosion.

Fig. 3.11 Mass loss of pure Mg after one week in solutions with different buffers

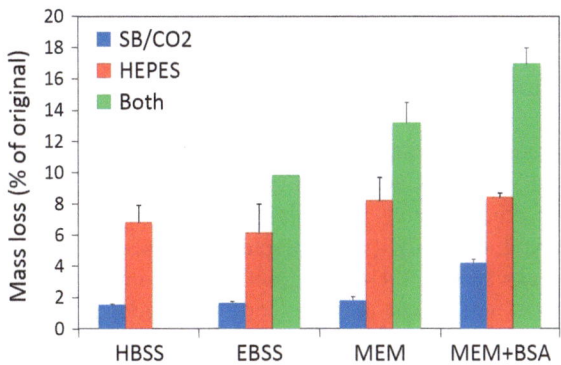

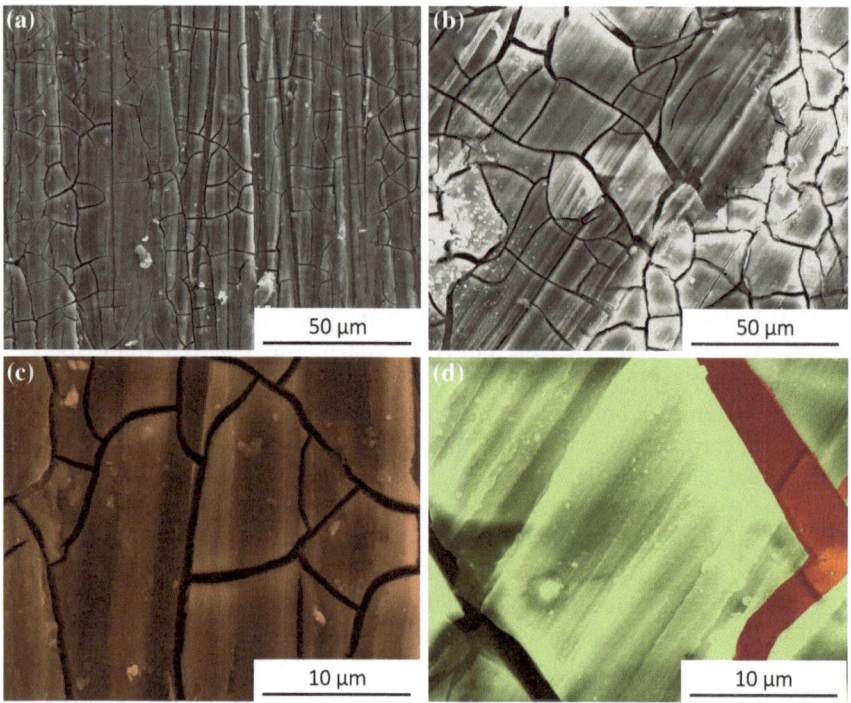

Fig. 3.12 Coloured scanning electron micrographs of pure Mg corroded for 3 h with HEPES **a, c** and SB_{CO_2} **a, a**. *Red* indicates $Mg(OH)_2$, *Green* indicates $CaP/CaCO_3$

3.3.7 Calcium/Magnesium Carbonate Formation

In addition to the various forms of CaP that form and offer protection to Mg, a potential reason for the decreased corrosion of samples in SB_{CO_2}-buffered media is the formation of calcium carbonate ($CaCO_3$) or magnesium carbonate ($MgCO_3$) on the surface. Calcium salts are known to be cathodic inhibitors of corrosion of metals [42], and $MgCO_3$ has also been previously shown to provide protection to pure Mg in a bicarbonate-buffered SBF [43]. Consequently, these carbonates may play a large role in the corrosion protection afforded to Mg in SB_{CO_2}-buffered media.

Analysis via Fourier transform infrared spectroscopy (FTIR) of the corrosion products of samples immersed in differently buffered media provided further evidence of $CaCO_3$ or $MgCO_3$ forming on the surface (Fig. 3.13). Carbonate absorbance peaks are typically found at 875–878 [44, 45] and 1,415 [46], 1,420, and 1,460 cm^{-1} [47]. Samples immersed in the SB_{CO_2} buffer system displayed peaks at 875, 1,416 and 1,460, confirming the presence of CO_3^{2-}. Similar peaks were not found for the corrosion products of samples buffered in HEPES.

Fig. 3.13 FTIR absorbance spectra of surfaces of pure Mg samples immersed with different buffers. Typical carbonate peaks are indicated

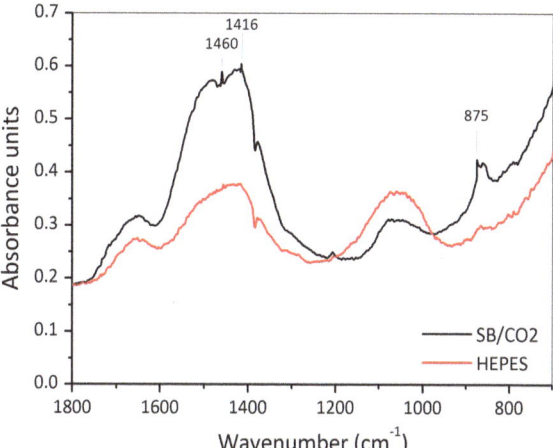

This supports the notion that $Ca/MgCO_3$ is providing some of the corrosion protection in the SB_{CO_2} buffer system.

Proteins may inhibit the formation of $CaCO_3$ during in vitro testing [48]. This may explain why the HEPES-buffered MEM_{BSA} and MEM displayed very similar results, while the SB_{CO_2}-buffered MEM_{BSA} displayed a twofold increase in corrosion.

To evaluate the effects of the HEPES buffer on Mg corrosion, independently of the formation of CaP layers on the Mg surface, experiments were performed in distilled water with and without HEPES buffer. It was found that the corrosion rate increased in distilled water in the presence of HEPES, which is likely due to two factors. Firstly, the buffer is hindering the formation and maintenance of a high-pH environment near the surface of the sample and thus also hindering the creation of a thicker, more protective corrosion layer. Secondly, the HEPES itself may be reacting with some component of the Mg substrate, increasing its corrosion.

This later concept has been explored extensively by the authors in [49], where the chemical interaction between Mg and HEPES was investigated using solution-state nuclear magnetic resonance (NMR). The results appeared to validate the assumptions that HEPES does not have a significant interaction with Mg^{2+} at levels typically seen in cell cultures [33]. However, it is likely that the local concentration of $MgCl_2$ near the surface of the corroding Mg is much higher than that of the general solution.

3.4 Choice of Simulated Body Fluid

Although to date there are a number of different solutions that have been used in the literature to investigate Mg biocorrosion, there have only been a few studies that have compared the corrosion behaviour in different solutions. Although these studies have explored some of the effects different media can have, they have been restricted to a very small number of alloys. In addition, the current literature suffers from a number of issues which limit its applicability: inappropriate use of surface analysis after polarisation [50], no pH control [51], or little discussion of results obtained [52]. Several other studies have also compared a few media, but all are again affected by the issues mentioned above [53–55].

3.4.1 High Chloride Content of Current SBFs

One component of SBFs that has been widely overlooked by the Mg biocorrosion community is their Cl^- content. The three most commonly used media, HBSS, EBSS, and MEM, all contain significantly more Cl^- than human plasma (HP) (Table 3.1). This is a result of the original purpose for which these media were designed. Virtually, all the solutions that have been used in the literature to investigate Mg biocorrosion were originally intended to be used for cell culture. The role of the salts in these culture media is partly to provide a physiological ionic environment for cell metabolism [29]. However, certain ions are considered more important than others. Sodium (Na) and potassium (K), for example, are essential for pump functions across the cell membrane and are crucial for cell survival and growth [29]. Therefore, in cell culture experiments, levels of these ions are kept as closely as possible to physiological conditions.

In all the SBFs used, the primary reagent is NaCl, which provides the bulk of the Na^+ and Cl^- in the solution. However, the amount of Cl^- does not appear to be as crucial to cell viability as Na^+ [29]. Na^+ can also be added to media as the compound Na_2HPO_4, but the amount of this compound is limited to less than 2 % of the total amount, because only a small amount of HPO_4^- is required. Thus, 7.5–8 g/l NaCl is typically added to SBF to obtain the desired Na^+ concentration. This in turn results in ~ 137 mmol/l of Cl^-— ~ 40 % more than that found in HP. Given the active role Cl^- is known to play in accelerating the corrosion of Mg alloys, this difference is probably the main reason why in vitro tests consistently report higher corrosion rates than those measured in vivo [56].

Table 3.1 Chloride contents of different SBF media	

Media	Cl^- content (mmol/l)
Human plasma	100–103
8 g/l NaCl	136.9
PBS	140
HBSS	144.6
EBSS	135
MEM	123.5

3.4.2 Design of a Biocorrosion Medium with Physiologically Correct Cl⁻ Levels (Kirkland's Medium)

The variance in Cl^- content shown in Table 3.1 highlights the fact that the majority of SBFs that have been used in the bio-Mg literature were not specifically designed for corrosion experiments. Although some researchers, including Kokubo [57, 58], Oyane [59] and Takadama [60], have created revised versions of a generic SBF with similar Cl^- concentrations to HP, these solutions contain less than 20 % (4.2 mmol/l) of the physiological level of HCO_3^-. The reason given for this low level was that physiological levels of HCO_3^- resulted in the deposition of calcite onto certain biomaterials because of the release of Ca^{2+} ions [60]. However, this occurs almost exclusively on Ca-rich biomaterials, such as hydroxyapatites [60]. It has been established that a minimum of 25 mmol/L of bicarbonate is necessary to buffer a typical SBF to 7.4, according to the Henderson–Hasselbalch equation [29]. This means that although these generic SBFs contain correct levels of Cl^-, they will always require the use of a chemical buffer.

The authors have developed a new modified SBF [Kirkland's biocorrosion medium (KBM), see Appendix A], with a lower chloride concentration, which also allows various buffers to be used.

3.4.3 Effect of Cl⁻ Content on Mass Loss

In an attempt to define the effect of the reduced Cl^- content of KBM on the corrosion rate of pure Mg, immersion tests were performed in both HBSS and KBM buffered with either HEPES or SB_{CO_2}. The change in Cl^- concentration appears to make a substantial difference to the total mass loss (Fig. 3.14). In both buffering environments, decreased corrosion was observed in KBM, with the

Fig. 3.14 Mass loss of pure Mg in KBM and HBSS

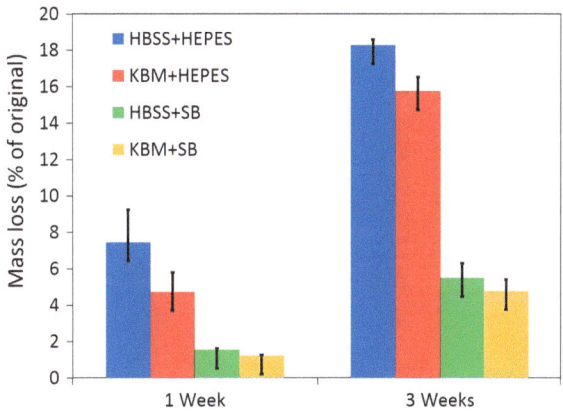

biggest difference being a $\sim$40 % decrease in mass loss after one week in HEPES-buffered media. Smaller differences (21 % and 13 % decreases at one and three weeks, respectively) occurred for samples buffered with SB_{CO_2}, possibly because of formation of a Mg carbonate layer on the surface in the SB_{CO_2} environment (see Sect. 3.3).

Changing the Cl^- concentration at the levels investigated alters the corrosion rate. This highlights the importance of the development and use of media (such as KBM) which more closely mimic the ionic environment of the body.

3.4.4 Mg Alloy Degradation as a Function of Organic Components

Proteins have been widely used in solutions for Mg biocorrosion studies, but their effects are poorly understood. To better understand the in vitro effect of proteins, corrosion of a wide range of alloys was investigated, in MEM with and without the addition of 10 % foetal bovine serum (FBS, a common protein-containing addition to cell cultures). Potentiodynamic polarisation experiments were used.

Proteins reduced the i_{corr} of each of 29 alloys investigated; not only was the measured current density decreased, but the variation between samples was also reduced (Fig. 3.15). As has been suggested elsewhere, this could indicate a more uniform layer on the surface resulting from rapid protein adsorption [52]. If we exclude the Mg-16.2Ca sample, the average i_{corr} for the alloys tested in MEM was 145 $\mu A/cm^2$, whereas the average for the MEM_{FBS} alloys was 52 $\mu A/cm^2$, almost 3 times lower than in MEM. A reduction in i_{corr} for Mg alloys in the presence of

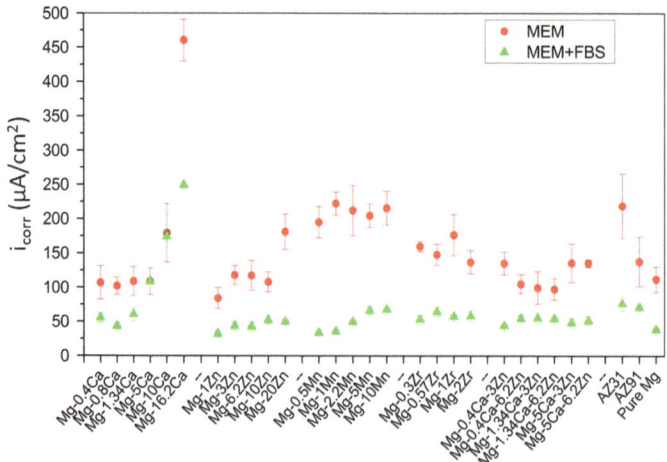

Fig. 3.15 i_{corr} for pure Mg and various binary and ternary alloy systems in MEM with and without 10 vol. % FBS

Fig. 3.16 i_{corr} of 13 alloys in PBS with 50 vol. % HP. MEM$_{FBS}$ data are included for comparison

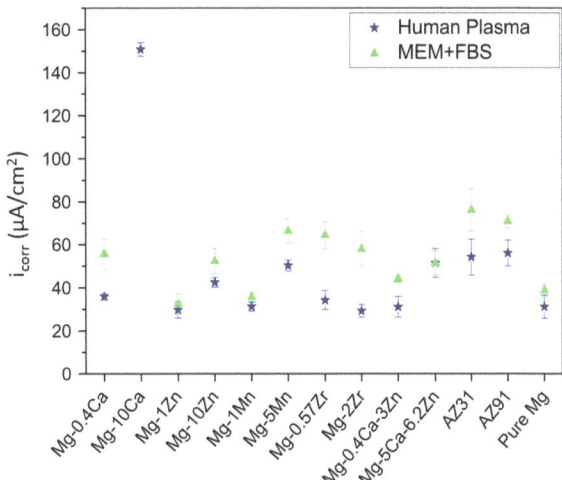

proteins has also been reported in the literature [43, 52, 61, 62]. It has been suggested that the proteins may: (1) thicken the Mg(OH)$_2$/CaP corrosion layer [43], (2) provide an undescribed "blocking effect" [52], (3) react rapidly with divalent ions, speeding the formation of a Mg(OH)$_2$/CaP corrosion layer [63], and (4) block the migration of Cl$^-$ ions to the surface [64, 65]. However, none of these reports provided further evidence to support these hypotheses. Although the reasons are at present unclear, there is general agreement that proteins significantly reduce the rates of the electrochemical reactions that are taking place.

To compare these values in a more physiological environment, alloys that displayed some of the lowest and highest i_{corr} values were chosen for further investigation in what is perhaps the most realistic in vitro solution used to date, PBS with 50 % HP (Fig. 3.16). HP is not commonly used to investigate biomaterials because it is difficult to handle, and because many materials have test standards in place that do not require such media. However, accepted test criteria do not exist for Mg, and we wished to investigate the alloys in conditions that were as realistic as possible in vitro.

The findings from the PDP tests were very interesting, with i_{corr} values in PBS$_{HP}$ 15–50 % lower than those in MEM$_{FBS}$, although most alloys display very similar results in MEM and PBS without added proteins. One possible reason for this difference is that the MEM$_{FBS}$ solution contains only 4–6 g/l of protein, whereas the HP has the full complement of 60–80 g/l of protein found in vivo. Thus, PBS with 50 vol. % HP contained 30–40 g/l total protein, up to 10 times more than MEM$_{FBS}$. The amount of protein that will adsorb to a surface is heavily dependent on the concentration in the solution [66]. Consequently, much more protein may be adsorbed onto the surface of the Mg alloys in PBS$_{HP}$ than in MEM$_{FBS}$, and these additional proteins may form a thicker, more protective layer, although this has yet to be confirmed.

The trends in electrochemical performance of a range of alloys in a 50 vol. % HP solution are relatively close to those in a similar calf-serum medium. Although more work is needed to confirm this behaviour, it appears that it may not be essential to use (potentially hazardous) human material in the search for a bio-realistic environment for Mg biodegradation.

3.4.5 Biocorrosion of Mg in Various In Vitro Media

As will be discussed in Chap. 4, chemical composition or microstructure (e.g. second phases and phase morphology) greatly influence the corrosion behaviour of Mg alloys. Consequently, one benefit of examining a single alloy in a range of solutions is that the number of variables that can cause discrepancies in the results is reduced. In this regard, it is advantageous to use pure Mg to ascertain the influence of media alone. A summary of the anodic and cathodic shifts due to the various media, and their effects on i_{corr} and E_{corr} is presented in Fig. 3.17.

Mg in KBM displayed a slight decrease in the rate of oxidation which, given the similarity of KBM to the other solutions, is likely to be because of the lower Cl⁻ content in KBM. PBS displayed a small negative cathodic shift, resulting in a decrease in E_{corr}, which may be because the different composition of the medium results in a different corrosion layer (explored further in Fig. 3.19). For the MEM media, increasing amounts of proteins resulted in a decrease in the reduction reaction rate, as shown by the negative cathodic shift. The addition of ∼4 wt. % proteins, from the FBS, resulted in the largest shift. Further increases in protein concentration made E_{corr} more electronegative, though by smaller amounts. The proteins appear to be mild cathodic inhibitors, decreasing the rate of H_2 evolution and the corrosion rate.

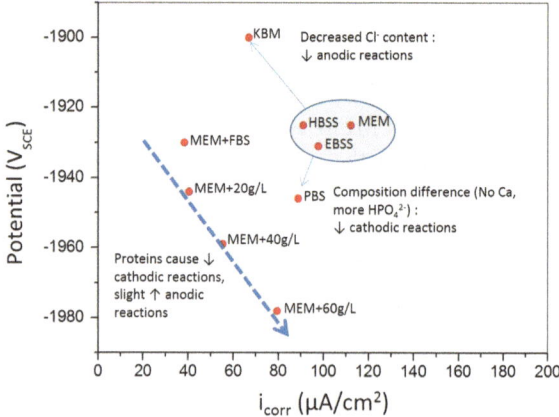

Fig. 3.17 E_{corr} versus i_{corr} (corrosion current density) for pure Mg a range of media with reaction shifts indicated

The i_{corr} values of Mg in the various media showed small but clear trends. Samples in KBM corroded approximately 30 % slower than those in HBSS, EBSS, and PBS, driven primarily by the decrease in the rate of anodic reactions. MEM displayed the greatest variation in i_{corr}; this suggests that although AAs play a role in the early stages of corrosion, they may be adhering to the surface inconsistently. The most interesting trend is that with the addition of proteins, where smaller amounts (such as in MEM with FBS or 20 g/L BSA) result in a decrease in i_{corr}, but higher concentrations lead to an increase (although the corrosion current in the presence of protein is always less than that in plain MEM). Combined with the observed decrease in the rate of cathodic reactions, this means that the proteins must be increasing the anodic reaction rate. This increase in anodic reaction rate in the presence of proteins has been previously reported for WE43 and LAE442 magnesium alloys [1]. Therefore, proteins slow the rate of corrosion of pure Mg primarily by limiting the cathodic reactions; increasing the amount of proteins to physiological levels in fact increases the anodic reaction rate.

3.4.6 EIS of MEM with Proteins: A Complicated System

MEM displayed similar polarisation behaviour to the other SBFs, with an initial peak in R_f observed at 3.5 h, after which the values settled (Fig. 3.18a). However, although MEM contains around 20 % less Cl^- than HBSS or EBSS, the R_{CT} values were significantly lower than those found in the other non-protein-containing SBFs. The inclusion of AAs may be the cause of this decrease; a number of reasons for this have been suggested in other studies (refer to Sect. 3.4.8).

The addition of proteins led to a number of interesting phenomena. BSA resulted in an initial peak in R_{CT} rather than R_f (as was seen for other solutions). However, this rise in R_{CT} did not last long—R_{CT} dropped to less than 50 % of its peak value after just 3 h of immersion. The R_{CT} of the 40 g/l and 60 g/l BSA solutions remained constant at $\sim$400–500 Ω/cm^2 until 30 h, after which they started to decrease. This may be because substantial corrosion had occurred, as shown in Fig. 3.19, which would have increased the surface area, further decreasing the resistance (which was determined based on the apparent surface area).

The surface film that formed in the presence of proteins did not provide longer-term protection, assessed from R_f (Fig. 3.18c). The formation of a CaP-containing layer was significantly hindered by the presence of proteins, with higher concentrations resulting in increasingly rapid degradation. This was further confirmed by EDS analysis (Table 3.2), and similar behaviour has been found for certain proteins in the literature [48]. After 30 h all protein-containing solutions displayed a single time constant, indicating that no protective layer remained. The lack of this semi-protective layer on the surface of the Mg is probably why the corrosion rate can increase in the presence of proteins.

Fig. 3.18 a R_{tot}, **b** R_{CT} and **c** R_f of pure Mg in MEM with varying amounts of BSA over 72 h. R_f is shown for only the first 30 h

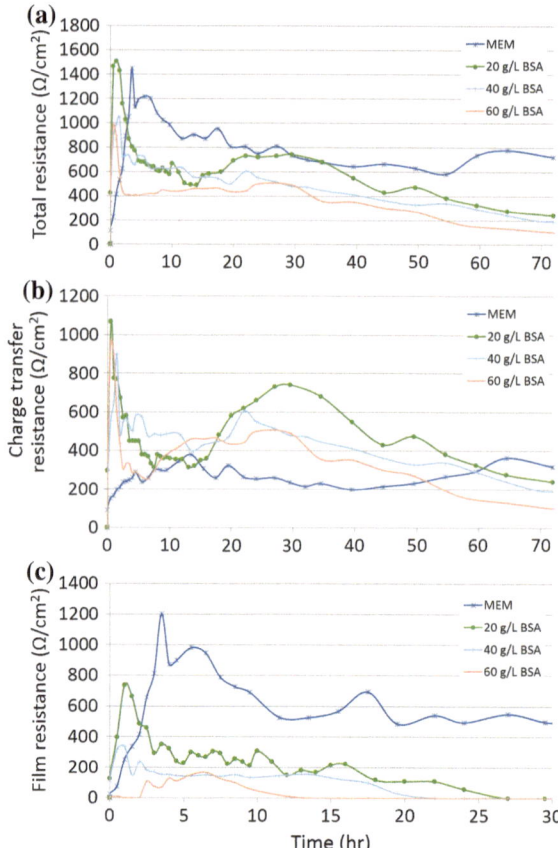

Table 3.2 Elemental composition (at % , according to EDS) of corrosion layer on pure Mg after testing in various solutions

Solution	Mg	O	Ca	P
NaCl	43.2	55.1	0	0
KBM	28.7	49.1	13.1	7.2
HBSS	29	48.5	10.4	9.8
EBSS	33.1	46.8	12.5	7.4
PBS	32.9	45	0.5	17
MEM	27	50.2	9.5	5.7
20 g/l	36.1	50.4	5.8	4.4
40 g/l	40.4	50.1	4.1	3.2
60 g/l	42.4	47.2	4.2	3

Values may not reach 100 % due to presence of other elements (Na, Cl) on the surface

3.4.7 Morphology and Composition of Corrosion Layers

The corrosion products formed in a NaCl medium showed a directionally cracked surface, where the directions of the cracks depended on the orientation of the underlying grains (Fig. 3.19a). Samples in KBM, HBSS, EBSS and PBS exhibited similar morphologies (Fig. 3.19b–e); a cracked-earth-like structure prevailed, similar to the one that is commonly found for Mg in SBFs [8, 50, 67, 68]. The layer had a mixed Ca–Mg-P composition as has been suggested previously [38]. Occasional pits were visible on the surface, usually surrounded by an area with no Ca or P. It has been suggested that the formation of Ca/P rich layers provides the bulk of the protection from corrosion for Mg alloys in most SBFs [43].

Samples corroded in MEM displayed more severely corroded morphologies, including cracks with sharper edges and peaks (Fig. 3.19f). The BSA-containing MEM solutions also displayed a fractured corrosion layer, but did not show the same crack pattern as the other solutions, instead displaying wider and deeper-looking gaps between the layers (Fig. 3.19g–i). For the protein-containing solutions, no obvious corrosion layer could be seen on the surface.

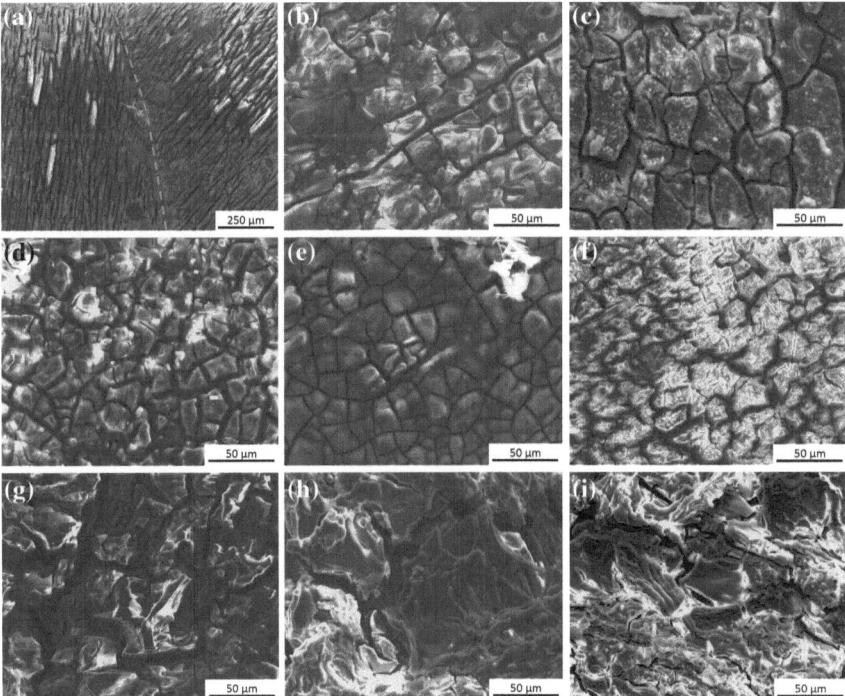

Fig. 3.19 Scanning electron micrographs of the surface of pure Mg samples after 72-h immersion in; **a** NaCl, **b** KBM, **c** HBSS, **d** EBSS, **e** PBS, **f** MEM, **g** MEM + 20 g/l BSA, **h** MEM + 40 g/l BSA, and **i** MEM + 60 g/l BSA. *Dotted line* in **a** indicates the grain boundary

EDS analysis of the surface of the sample immersed in the NaCl solution found only Mg and O (Table 3.2). This could indicate that a layer of $Mg(OH)_2$ was present, as is expected for Mg in saline solutions [38]. The surfaces after exposure to KBM, HBSS and EBSS showed similar compositions—an even distribution of Mg, O, Ca, and P. The surfaces of samples immersed in PBS displayed high amounts of P and small amounts of Ca. There may be two reasons for this: firstly, PBS media does not contain Ca^{2+}, so there should be very little Ca^{2+} available in the solution to form compounds and precipitate onto the surface [69], and secondly, Rettig and Virtanen found that magnesium phosphate tribasic $[Mg_3(PO_4)_2]$ will readily form in solutions without Ca, but Ca will not be deposited on a Mg alloy surface without HPO_4^{2-} [38]. Therefore, it is likely that the surfaces of the samples immersed in PBS were a combination of $Mg(OH)_2$ and $Mg_3(PO_4)_2$, as indicated by EDS (Table 3.2).

In the MEM media, as the concentration of BSA increased, the amounts of Ca and P in the corrosion layer decreased. The surface layers produced on samples immersed in MEM with 60 g/L BSA contained less than half the concentration of Ca and P of the surface layers immersed in KBM, HBSS and EBSS. Although the thickness of the corrosion layers could not be determined from the micrographs (Fig. 3.19), EDS indicated that the Mg content of the layers increased as the protein concentration rose. However, the concentration of oxygen on the surface remained relatively stable, irrespective of the BSA addition. The increased concentration of Mg may be because the corrosion layer is thin enough that the EDS detected the pure Mg sublayer.

> The use of EDS to determine the presence of proteins by detection of C and O is speculative, because these elements could come from other sources (e.g. $Mg(OH)_2$, or Mg and Ca carbonates). Although the detection of C and O on the surface to confirm protein attachment has been reported, this is not fully justified [43, 70]. It has been suggested that it might be possible to detect the nitrogen (N) that is within the AA to indirectly measure the presence of proteins [38], however EDS analysis is not suited for measurement of nitrogen.

3.4.8 Importance of Amino Acids

AAs are molecules containing C, H, O and N which perform many roles in the body; they are vital nutrients and form the building blocks for proteins [71]. They are also known to affect corrosion, as several studies on aluminium alloys have found significant corrosion inhibition in the presence of AA [72–74], while others have found similar effects for steel [75] and lead alloys [76]. Recently, a few studies have indirectly looked at the effect of AA on the performance of Mg alloys in vitro. Some studies have found corrosion to increase; suggested reasons for this include the AA forming a complex with Mg^{2+} ions [77, 78] or inhibiting the formation of the $Mg(OH)_2$/CaP layer via chelation, where metal ions attach to a large molecule and are removed from the surface [43]. However, other studies observed AA to result in a decrease in the corrosion rate of Mg alloys [50]; proposed reasons for this included the AAs acting as cathodic inhibitors [72, 75, 76] or limiting the presence of pitting corrosion by inducing potential shifts [73, 74].

In this work, AA have been shown to affect Mg in vitro by altering the morphology (Fig. 3.19) and elemental composition (Table 3.2) of the corrosion layer. AA also appear to have an effect on the electrochemical corrosion performance of pure Mg, changing the way in which the surface films form as well as lowering the overall resistance (Fig. 3.18). However, how AA affect Mg biodegradation in vivo remains unknown, and this area warrants further study.

3.4.9 Influence of Cellular Attachment on Mg Biodegradation

Cells are another addition to SBFs that may be employed in the drive to create biorealistic solutions. Cellular attachment is critical to the success of an Mg implant and may play a considerable role in its corrosion. For example, macrophages and the active oxygen species they generate are reported to accelerate corrosion of Ti [79]. Although cells were not investigated as part of this work, relevant literature has been collected to provide a discussion in the light of the presented work.

In the bio-Mg literature, the vast majority of experiments that have used cells have been toxicity studies on various alloys [67, 80–91]. While most have found that the alloys have caused little toxicity for the cells, they generally have not investigated the effect cellular attachment can have on corrosion.

Recently, several studies have made initial steps to investigate this. Feser et al. looked at the concentration of Mg (mmol/l) in cell media for a range of Mg–Ca alloys (0.6–1.2 wt % Ca) [92]. They found increasing levels of Mg as the Ca content increased. However, little discussion was provided regarding the mechanism or reasons for differences in corrosion. Wong et al. compared cell viability with Mg^{2+} ion release over time, but did not directly look at the effect of cells on the corrosion [93]. In perhaps the most relevant study, Gu et al. examined the Mg, Ca and Zn concentrations at 1, 3 and 5 days for a range of Mg bulk metallic glasses in DMEM with and without L929 or MG63 cells [41]. It was found that several alloys corroded around 30–50 % slower in media containing cells; however, pure Mg remained almost unchanged. They discussed the effects that H_2 evolution, rapid ion release and pH changes may have on cell attachment.

In related work, Hiromoto et al. described how cells may slow the interaction of Cl^- ions with the surface of metal implants, reducing corrosion significantly [94].

Witte et al. also reported that giant cells have actually engulfed small amounts of Mg corrosion product, although this study was more closely related to toxicity [95].

It is clear that cellular activity is a crucial part of the success of a Mg implant in vivo. However, the potential roles that cells may play in both the short- and long-term degradation mechanisms of Mg alloys remain to be established.

3.5 Effects of Sample Preparation and Surface Roughness

To allow realistic and meaningful comparisons between experimental results, it is crucial to consider every step taken in the preparation of samples. Preparation steps that can influence the corrosion rate include the following:

1. Cutting of samples—may cause subsurface deformation that changes micro-structure, leaving residual stresses that may go deep (>50 μm) into the surface [96, 97],
2. Cleaning—contaminants left on surfaces of samples can affect corrosion and/or biocompatibility, particularly iron pickup from tooling!
3. Mounting—it has been suggested [98] that mounting a sample can cause increased crevice corrosion at the interface of the sample and mounting material; however, this is most likely due to contamination at cut edges (related to item 2 above) since the corrosion of Mg should be independent of oxygen concentration,
4. Grinding/polishing—different roughness and surface topology may affect both corrosion [99] and protein/cell adhesion [100].

The most easily controlled of these parameters is the grinding or polishing of the samples to a desired roughness (R_a). Directly after casting, the surface topology may vary widely depending on location in the cast mould or ingot. Grinding can help normalise the surfaces of the samples and may also be used to remove any deep residual stresses arising from prior deformation. Changes in surface roughness will also affect the active surface area; rougher surfaces with deeper "valleys" display greater area. This in turn can affect both immersion and electrochemical experiments, where surface area must be known to calculate corrosion parameters [101].

> Grinding is effectively a local form of severe plastic deformation and may alter more than just the surface roughness of the material [102]. Grinding may cause local changes in composition (redistribution of alloying elements or segregation of impurities), residual strain and lead to embedding of foreign particles (such as SiC) [103].

To investigate this, samples of pure Mg were ground/polished to a variety of theoretical roughness grades, after which their actual roughness value was measured using a stylus profilometer (Table 3.3).

Table 3.3 Theoretical and measured surface roughness of pure Mg polished to various grades

Nominal Polish	Theoretical Roughness (μm)	Measured Roughness (μm)
180 grit	63	23.1 ± 5.77
600 grit	20	4.4 ± 1.3
1,200 grit	9.5	2.2 ± 0.2
1 μm	1	1.2 ± 0.1
0.02 μm	0.02	0.04 ± 0.04

EIS elucidated some interesting behaviour related to ΔR_a (Fig. 3.20). The highly polished, 0.02 μm samples displayed a single time constant, with no appreciable resistance due to film formation. The 9.5 μm samples exhibited the characteristic R_f peak after a few hours, as did the 20 μm samples, although to a lesser extent. However, the 63 μm samples did not display this peak, instead R_f rose slowly to 400 Ω/cm^2, almost exactly the same value as found for the 20 μm samples (Fig. 3.20c).

Fig. 3.20   a R_{tot}, **b**, R_{CT}, **c** R_f of pure Mg as a function of surface roughness over 72 h

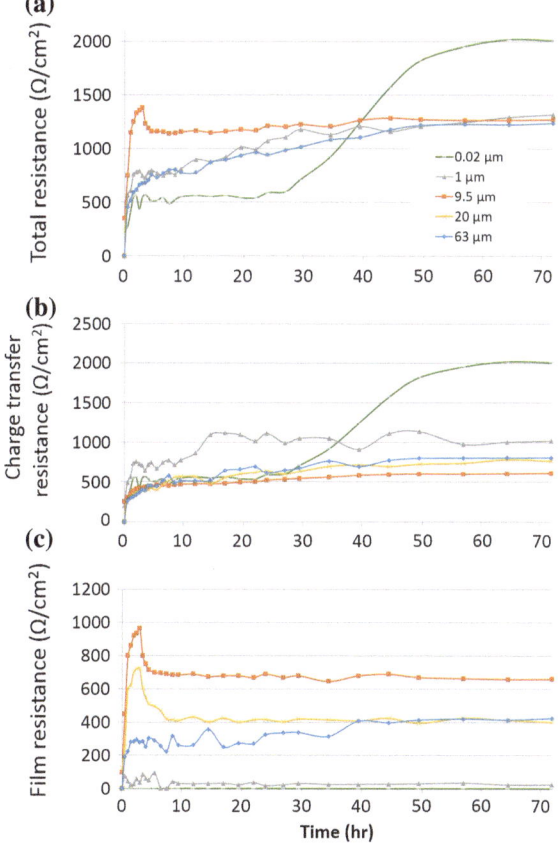

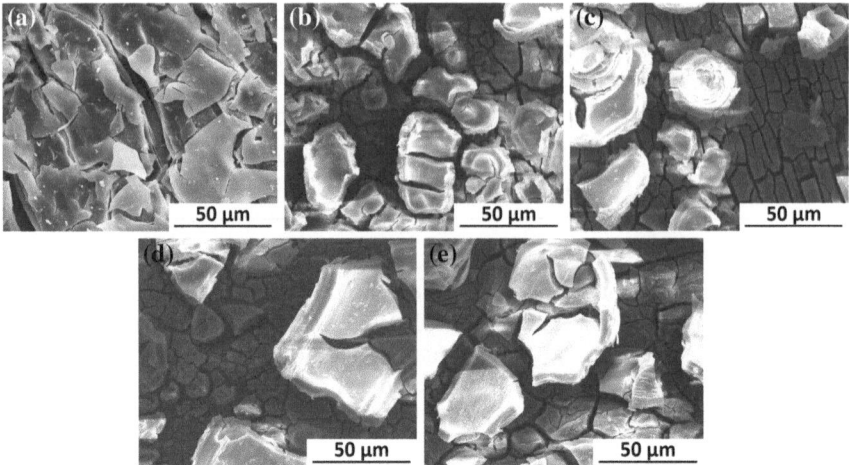

Fig. 3.21 Scanning electron micrographs of the surface of pure Mg samples after 72-h immersion as a function of surface roughness; **a** 0.02 μm, **b** 1 μm, **c** 9.5 μm, **d** 20 μm, and **e** 62 μm

The R_{CT} displayed a similar trend for the four rougher surfaces, although the 1 μm surface exhibited consistently greater resistance. However, 0.02 μm samples demonstrated radically different behaviour, with R_{CT} rising sharply after 30 h to more than double that of the three roughest samples. This difference is due to a different surface morphology, as examined by SEM (Fig. 3.21).

All samples, other than the 0.02 μm, displayed strikingly similar R_{tot} values, within 100 Ω/cm^2 of each other (Fig. 3.20a). This indicates that although the differences in surface preparation resulted in varied corrosion in the initial hours of immersion, the effect was minimal after 2 days. However, the smoothest surface displayed significantly altered corrosion behaviour, with approximately 60 % greater R_{tot} over the 72 h.

3.5.1 Analysis of Corroded Surfaces

The surfaces of the samples provided some indication of how the corrosion proceeded (Fig. 3.21). All surface finishes other than the 0.02 μm resulted in a similar corrosion layer, with a $Mg(OH)_2$ "cracked-earth" sublayer and CaP flakes above. This surface morphology is commonly observed for pure Mg in balanced salt media, both in this work and the bio-Mg literature [8, 50, 67, 68].

However, the 0.02 μm samples displayed a more amorphous, smooth layer which contained similar levels of Ca and P as the flakes on the other samples (Fig. 3.21a). The coverage of this layer was also far greater than on the other samples; the layer appeared to cover the entire surface. It is probable that the surface would be completely covered by this layer during immersion in HBSS. This layer may have resulted in the significant increase in corrosion protection that was shown in the EIS data (Fig. 3.20). In addition, the complete coverage and protection that is provided is consistent with the single time constant in the EIS of the 0.02 μm samples.

3.6 Summary

Temperature has a major influence on the corrosion rate of a range of Mg alloys, with higher temperature leading to significantly faster degradation (Fig. 3.3). The extent of the increase in corrosion is dependent on both the alloy and the solution, making it difficult to estimate the increase for a given alloy/solution combination without prior experimentation. It is crucial, therefore, that any experiment be carried out at the normal T_{phys}, 37 °C.

For Mg in SBF, an apparently small variation in pH, from 7.2 to 7.8, results in a large decrease in corrosion rate (220–300 %) (Fig. 3.5). Such "minor" variances in pH are within the critical formation range for the precipitation of CaP in SBF (Fig. 3.10), which may result in radically different corrosion morphologies on the surface (Fig. 3.8). pH also affects other surface interactions, including interactions with AAs and protein adhesion, that will further alter biodegradation behaviour. Consequently, adjustment and maintenance of the pH to physiological levels (7.4–7.6) throughout the entire experiment is critical.

An unbuffered solution (especially in the vicinity of the metal surface) will often display a rapid increase in pH, because of the release of OH^- during Mg degradation. This will then create an unphysiological environment for the remainder of the experiment, dramatically changing the degradation process. The use of HEPES and other chemical buffers provides one method of controlling the pH for in vitro tests, although these buffers are different to the way the body regulates pH. Ideally, biocorrosion testing of Mg should be carried out using a physiological amount of $NaHCO_3$ in a controlled CO_2 atmosphere (i.e. the closest in vitro condition to the human body). Regardless, the use of a buffering agent is, for all but the shortest experiments, a vital part of experimental design.

The correct choice of an appropriate medium is essential in the quest to obtain valuable data. In the light of this, NaCl-only solutions cannot be considered suitable for the investigation of Mg biodegradation, as the complete lack of other inorganic salts results in considerable differences in both corrosion rates and mechanisms. Balanced salt solutions (i.e. HBSS, EBSS) provide a more suitable inorganic environment, although care must be taken to ensure they contain the correct physiological concentrations of those elements that significantly impact the degradation process, including Cl^-, Ca and P. AAs and proteins increase the physiological relevance of these solutions, but a much greater understanding of the underlying corrosion mechanisms is needed before they may be confidently and correctly employed in vitro.

The roughness of the sample surface needs to be controlled throughout any group of tests to allow for valid comparison of results. Changing the surface roughness resulted in different corrosion rates, although the overall degradation mechanism for typical polishing techniques appeared comparable (with the exception of very finely polished Mg). Consequently, a finer polish and lower surface roughness do not necessarily guarantee slower in vitro corrosion. While no definite recommendation for specific polishing prior to testing can be made at this time, a medium grade, such as 1,200 grit (9.5 µm), appears to be best; polishing to this grade allows for initial polishing stages to help remove deep residual stresses from the surface and finer papers to provide a consistent surface topology.

References

1. Mueller WD, Lucia Nascimento M et al (2010) Critical discussion of the results from different corrosion studies of Mg and Mg alloys for biomaterial applications. Acta Biomater 6(5):1749–1755
2. Kim W-C, Kim J-G et al (2008) Influence of Ca on the corrosion properties of magnesium for biomaterials. Mater Lett 62(25):4146–4148
3. Denkena B, Lucas A (2007) Biocompatible magnesium alloys as absorbable implant materials—adjusted surface and subsurface properties by machining processes. CIRP Ann: Manuf Technol 56(1):113–116
4. Hassel T, Bach FW et al (2006) Investigation of the mechanical properties and the corrosion behaviour of low alloyed magnesium–calcium alloys for use as absorbable biomaterial in the implant technique. In: Pekguleryuz M (ed) Conference of metallurgists: magnesium technology in the global age, Montreal, Quebec, Canada, pp 359–369
5. Lee J-Y, Han G et al (2009) Effects of impurities on the biodegradation behavior of pure magnesium. Met Mater Int 15(6):955–961
6. Peng Q, Huang Y et al (2010) Preparation and properties of high purity Mg-Y biomaterials. Biomaterials 31(3):398–403
7. Lopez HY, Cortes DA et al (2006) In vitro bioactivity assessment of metallic magnesium. Key Eng Mater 309(311):453–456
8. Yang L, Zhang E (2009) Biocorrosion behavior of magnesium alloy in different simulated fluids for biomedical application. Mater Sci Eng, C 29(5):1691–1696
9. Jones DA (1992) Principles and prevention of corrosion. Prentice-Hall, Englewood Cliffs
10. Burstein GT, Liu C (2007) Nucleation of corrosion pits in Ringer's solution containing bovine serum. Corros Sci 49(11):4296–4306
11. Omanovic S, Roscoe SG (1999) Electrochemical studies of the adsorption behavior of bovine serum albumin on stainless steel. Langmuir 15(23):8315–8321
12. Trépanier C, Pelton AR (2004) Effect of temperature and pH on the corrosion resistance of passivated nitinol and stainless steel. In: Proceedings of international conference on shape memory and superelastic technologies, ASM International, Baden, Germany, pp 361–366
13. Burstein GT, Liu C et al (2005) The effect of temperature on the nucleation of corrosion pits on titanium in Ringer's physiological solution. Biomaterials 26(3):245–256
14. Cavanaugh MK, Birbilis N et al (2012) Modeling pit initiation rate as a function of environment for Aluminium alloy 7075-T651. Electrochim Acta 59:336–345
15. Lambert RA (1913) The influence of temperature and fluid medium on the survival of embryonic tissues in vitro. J Exp Methods 18(4):406–411
16. Kirkland NT, Staiger MP et al (2011) Performance-driven design of biocompatible Mg-alloys. JOM 63(6):28–34

17. Gerasimov VV, Rozenfeld IL (1957) Effect of temperature on the rate of corrosion of metals. Russ Chem Bull 6(10):1192–1197
18. Makar GLJK (1993) Corrosion of magnesium. Int Mater Rev 38(3):138–153
19. Song G, Atrens A et al (1997) The anodic dissolution of magnesium in chloride and sulphate solutions. Corros Sci 39(10–11):1981–2004
20. Duygulu O, Kaya RA et al (2007) Investigation on the Potential of Magnesium Alloy AZ31 as a Bone Implant. Mater Sci Forum 546–549:421–424
21. Layrolle P, Daculsi G (2009) Physiochemistry of apatite and its related calcium phosphates. In: Leon B, Jansen JA (eds) Thin calcium phosphate coatings for medical implants. Springer, New York
22. Yin G, Liu Z et al (2002) Impacts of the surface charge property on protein adsorption on hydroxyapatite. Chem Eng J 87(2):181–186
23. Sharpe JR, Sammons RL et al (1997) Effect of pH on protein adsorption to hydroxyapatite and tricalcium phosphate ceramics. Biomaterials 18(6):471–476
24. Liang H, Huang F et al (2007) Enhanced calcium phosphate precipitation on the surface of Mg-ion-implanted ZrO_2 bioceramic. Surf Rev Lett 14(1):71–77
25. Lu X, Leng Y (2005) Theoretical analysis of calcium phosphate precipitation in simulated body fluid. Biomaterials 26(10):1097–1108
26. Ng WF, Chiu KY et al (2010) Effect of pH on the in vitro corrosion rate of magnesium degradable implant material. Mater Sci Eng C 30(6):898–903
27. Waters JH, Miller LR et al (1999) Cause of metabolic acidosis in prolonged surgery. Crit Care Med 27(10):2142–2146
28. Chakkalakal DA, Mashoof AA et al (1994) Mineralization and pH relationships in healing skeletal defects grafted with demineralized bone matrix. J Biomed Mater Res 28(12):1439–1443
29. Malda J, Woodfield TBF et al (2008) Cell nutrition: in vitro and in vivo. Tissue Eng: A Textbook 1:327–362
30. Hall JE (2010) Guyton and hall textbook of medical physiology, 11th edn. Elsevier, Amsterdam
31. Boron WF, Boulpaep EL (eds) (2008) Medical physiology, 2nd edn. Saunders, New York
32. Sigma-Aldrich (2010) Fundamental techniques in cell culture: A laboratory handbook. Sigma-Aldrich. http://www.sigmaaldrich.com/life-science/cell-culture/learning-center/ecacc-handbook/cell-culture-techniques-6.html#Buffering. Accessed 8 Oct 2010
33. Good NE, Winget GD et al (1966) Hydrogen ion buffers for biological research. Biochemistry 5(2):467–477
34. Masters JRW (ed) (2000) Animal cell cultures. Oxford University Press, Oxford
35. Eley JH (1988) The use of hepes as a buffer for the growth of the cyanobacterium Anacystis nidulans. Appl Microbiol Biotechnol 28(3):297–300
36. Yang JX, Cui FZ et al (2009) Characterization and degradation study of calcium phosphate coating on magnesium alloy bone implant in vitro. IEEE Trans Plasma Sci 37(7):1161–1168
37. Puigdomenech I (2013) Medusa chemical equilibrium calculator. Royal Institute of Technology, Sweden
38. Rettig R, Virtanen S (2009) Composition of corrosion layers on a magnesium rare-earth alloy in simulated body fluids. J Biomed Mater Res: Part A 88(2):359–369
39. Sigma-Aldrich (2010) RPMI-1640 medium: Dutch modification. Sigma-Aldrich Inc. http://www.sigmaaldrich.com/. Accessed 8 Oct 2010
40. Montemor MF, Simões AM et al (2007) Characterization of rare-earth conversion films formed on the AZ31 magnesium alloy and its relation with corrosion protection. Appl Surf Sci 253(16):6922–6931
41. Gu X, Zheng Y et al (2010) Corrosion of, and cellular responses to Mg-Zn-Ca bulk metallic glasses. Biomaterials 31(6):1093–1103
42. Roberge PR (2000) Handbook of corrosion engineering. McGraw-Hill, New York
43. Yamamoto A, Hiromoto S (2009) Effect of inorganic salts, amino acids and proteins on the degradation of pure magnesium in vitro. Mater Sci Eng C 29(5):1559–1568

44. Regnier P, Lasaga AC et al (1994) Mechanism of CO_3^{2-} substitution in carbonate-fluorapatite: evidence from FTIR Spectroscopy, ^{13}C NMR, and quantum mechanical calculations. Am Mineral 79(9–10):809–818

45. Rey C, Collins B et al (1989) The carbonate environment in bone mineral: a resolution-enhanced fourier transform infrared spectroscopy study. Calcif Tissue Int 45(3):157–164

46. Tatzber M, Stemmer M et al (2007) An alternative method to measure carbonate in soils by FT-IR spectroscopy. Environ Chem Lett 5(1):9–12

47. Doi Y, Moriwaki Y et al (1982) ESR and IR studies of carbonate-containing hydroxyapatites. Calcif Tissue Int 34(1):178–181

48. Burgess SK, Carey DM et al (1992) Novel protein inhibits in vitro precipitation of calcium carbonate. Arch Biochem Biophys 297(2):383–387

49. Kirkland N, Waterman J et al (2012) Buffer-regulated biocorrosion of pure magnesium. J Mater Sci Mater Med 23(2):283–291

50. Gu XN, Zheng YF et al (2009) Influence of artificial biological fluid composition on the biocorrosion of potential orthopedic Mg-Ca, AZ31, AZ91 alloys. Biomed Mater 4(6):8

51. Xin Y, Hu T et al (2010) Influence of test solutions on in vitro studies of biomedical magnesium alloys. J Electrochem Soc 157(7):C238–C243

52. Liu C, Xin Y et al (2007) Degradation susceptibility of surgical magnesium alloy in artificial biological fluid containing albumin. J Mater Res 22(7):1806–1814

53. Eliezer A, Witte F (2010) Corrosion behaviour of magnesium alloys in biomedical environments. Adv Mater Res 95:17–20

54. Liu C, Xin Y et al (2007) Influence of heat treatment on degradation behavior of bio-degradable die-cast AZ63 magnesium alloy in simulated body fluid. Mater Sci Eng, A 456(1–2):350–357

55. Alvarez-Lopez M, Pereda MD et al (2010) Corrosion behaviour of AZ31 magnesium alloy with different grain sizes in simulated biological fluids. Acta Biomater 6(5):1763–1771

56. Witte F, Nellesen J et al (2006) In vitro and in vivo corrosion measurements of magnesium alloys. Biomaterials 27(7):1013–1018

57. Kokubo T, Kushitani H et al (1990) Solutions able to reproduce in vivo surface-structure changes in bioactive glass-ceramic. J Biomed Mater Res 24:721–734

58. Kokubo T, Takadama H (2006) How useful is SBF in predicting in vivo bone bioactivity? Biomaterials 27(15):2907–2915

59. Oyane A, Kim H-M et al (2003) Preparation and assessment of revised simulated body fluids. J Biomed Mater Res: Part A 65A(2):188–195

60. Takadama H, Hashimoto M et al (2004) Round-robin test of SBF for in vitro measurement of apatite-forming ability of synthetic materials. Phosphorus Res Bull 17:119–125

61. Liu CL, Zhang XM et al (2010) In vitro corrosion degradation behaviour of Mg-Ca alloy in the presence of albumin. Corros Sci 52(10):3341–3347

62. Mueller WD, Nascimento ML et al (2007) Magnesium and its alloys as degradable biomaterials: corrosion studies using potentiodynamic and EIS electrochemical techniques. Mater Res 10:5–10

63. Klinger A, Steinberg D et al (1997) Mechanism of adsorption of human albumin to titanium in vitro. J Biomed Mater Res 36:387–392

64. Vogt C, Bechstein K et al (2008) Investigation of the degradation of biodegradable Mg implant alloys in vitro and in vivo by analytical methods. In: Kainer KU (ed) Proceedings of 8th international conference on magnesium alloys and their applications, Weimar, Germany, Wiley-VCH, pp 1162–1174

65. Padilla N, Bronson A (2007) Electrochemical characterization of albumin protein on Ti-6al-4v alloy immersed in a simulated plasma solution. J Biomed Mater Res Part A 81A(3):531–543

66. Langmuir I (1916) The constitution and fundamental properties of solids and liquids. J Am Chem Soc 38:2221–2295

67. Xu L, Pan F et al (2009) In vitro and in vivo evaluation of the surface bioactivity of a calcium phosphate coated magnesium alloy. Biomaterials 30(8):1512–1523

68. Gu XN (2010) Microstructure, biocorrosion and cytotoxicity evaluations of rapid solidified Mg–3Ca alloy ribbons as a biodegradable material. Biomed Mater 5(3):035013
69. León B, Jansen JA (eds) (2008) Thin calcium phosphate coatings for medical implants. Springer, New York
70. Mueller WD, de Mele MFL et al (2009) Degradation of magnesium and its alloys: dependence on the composition of the synthetic biological media. J Biomed Mater Res: Part A 90A(2):487–495
71. Kohrer C, Bhandary UR (2009) Protein engineering. Nucleic acids and molecular biology, vol 22. Springer, Berlin
72. Ashassi-Sorkhabi H, Ghasemi Z et al (2005) The inhibition effect of some amino acids towards the corrosion of Aluminium in 1 m Hcl+ 1 m H_2SO_4 Solution. Appl Surf Sci 249(1–4):408–418
73. El-Shafei AA, Moussa MNH et al (1997) Inhibitory effect of amino acids on al pitting corrosion in 0.1 M NaCl. J Appl Electrochem 27(9):1075–1078
74. Bereket G, Yurt A (2001) The inhibition effect of amino acids and hydroxy carboxylic acids on pitting corrosion of Aluminium alloy 7075. Corros Sci 43(6):1179–1195
75. Ashassi-Sorkhabi H, Majidi MR et al (2004) Investigation of inhibition effect of some amino acids against steel corrosion in HCl solution. Appl Surf Sci 225(1–4):176–185
76. Kiani MA, Mousavi MF et al (2008) Inhibitory effect of some amino acids on corrosion of Pb-Ca-Sn alloy in sulfuric acid solution. Corros Sci 50(4):1035–1045
77. William DF, William RL (2004) Degradative effects of the biological environment on metals and ceramics. In: Ratner BD, Hoffman AS, Schoen FJ, Lemons JE (eds) Biomaterials science: an introduction to materials in medicine. Elsevier Academic Press, San Diego, p 430
78. Bruneel N, Helsen JA (1988) In vitro simulation of biocompatibility of Ti-Al-V. J Biomed Mater Res 22(3):203–214
79. Mu Y, Kobayashi T et al (2000) Metal ion release from titanium with active oxygen species generated by rat macrophages in vitro. J Biomed Mater Res 49(2):238–243
80. Li Z, Gu X et al (2008) The development of binary Mg-Ca alloys for use as biodegradable materials within bone. Biomaterials 29(10):1329–1344
81. Zhang S, Zhang X et al (2010) Research of Mg-Zn alloy as degradable biomaterial. Acta Biomater 6(2):626–640
82. Ren Y, Wang H et al (2007) Study of biodegradation of pure magnesium. Key Eng Mater 342–343:601–604
83. Witte F, Feyerabend F et al (2007) Biodegradable magnesium-hydroxyapatite metal matrix composites. Biomaterials 28(13):2163–2174
84. Lorenz C, Brunner JG et al (2009) Effect of surface pre-treatments on biocompatibility of magnesium. Acta Biomater 5(7):2783–2789
85. Zhang S, Li J et al (2009) In vitro degradation, hemolysis and MC3T3-E1 cell adhesion of biodegradable Mg-Zn alloy. Mater Sci Eng C 29(6):1907–1912
86. Zhang E, Yin D et al (2009) Microstructure, mechanical and corrosion properties and biocompatibility of Mg-Zn-Mn alloys for biomedical application. Mater Sci Eng C 29(3):987–993
87. Witte F, Feyerabend F et al (2006) Unphysiologically high magnesium concentrations support chondrocyte proliferation and redifferentiation. Tissue Eng 12(12):3545–3556
88. Pietak AM, Mahoney T et al (2007) Bone-like matrix formation on magnesium and magnesium alloys. J Biomed Mater Res 19(1):407–415
89. Gu X, Zheng Y et al (2009) In vitro corrosion and biocompatibility of binary magnesium alloys. Biomaterials 30(4):484–498
90. Feyerabend F, Fischer J et al (2010) Evaluation of short-term effects of rare earth and other elements used in magnesium alloys on primary cells and cell lines. Acta Biomater 6(5):1834–1842
91. Yun Y, Dong Z et al (2009) Biodegradable Mg corrosion and osteoblast cell culture studies. Mater Sci Eng C 29(6):1814–1821

92. Feser K, Kietzmann M et al (2010) Effects of degradable Mg-Ca alloys on dendritic cell function. J Biomater Appl 25(7):685–697
93. Wong HM, Yeung KWK et al (2010) A biodegradable polymer-based coating to control the performance of magnesium alloy orthopaedic implants. Biomaterials 31:2084–2096
94. Hiromoto S (2008) Corrosion of metallic biomaterials in cell culture environments. Electrochem Soc Interface 17:41–44
95. Witte F, Ulrich H et al (2007) Biodegradable magnesium scaffolds: part 1: appropriate inflammatory response. J Biomed Mater Res: Part A 81:748–756
96. Love LC (1985) Principles of metallurgy. Reston Publishing Company, Reston
97. Doege E, Droder K (2003) Deformation of magnesium. In: Kainer KU (ed) Magnesium—alloys and technologies. Wilkey-VCH Verlag GmbH, Weinheim
98. Shi Z, Atrens A (2011) An innovative specimen configuration for the study of Mg corrosion. Corros Sci 53(1):226–246
99. Alvarez RB, Martin HJ et al (2010) Corrosion relationships as a function of time and surface roughness on a structural AE44 magnesium alloy. Corros Sci 52(5):1635–1648
100. Gentile F, Tirinato L et al (2010) Cells preferentially grow on rough substrates. Biomaterials 31(28):7205–7212
101. Bruckenstein S, Sharkey JW et al (1985) Effect of polishing with different size abrasives on the current response at a rotating disk electrode. Anal Chem 57(1):368–371
102. Samuels LE (2003) Metallographic polishing by mechanical methods, 4th edn. ASM International, Materials Park
103. Gale WF, Totemeir TC (eds) (2004) Smithells metals reference book, 8th edn. Elsevier Inc., Oxford

Chapter 4
Developments in Mg-based Alloys for Biomaterials

Abstract The high susceptibility to corrosion of Mg and its alloys is an important aspect of its use as a biomaterial. Aside from Mg sacrificial anodes, the only practical application where corrosion of Mg is a requirement is bioresorbable metals. In this chapter, we introduce the relevant metallurgy of Mg alloys and relate metallurgy to biocorrosion rates. We describe developments in Mg alloys that are relevant to bioresorbable metals and also describe Mg-based metallic glasses.

Keywords Magnesium · Metallurgy · Metal · Metallic glass · Alloys · Alloying element · BMG · Toxicity · Coating

4.1 Introduction

4.1.1 Mg and Its Alloys

Magnesium is very widely available, with approximately 1 million tonnes per annum produced; 80 % of production comes from China [1]. Commercial purity now readily exceeds 99.8 %, a property essential for corrosion control. As far back as 1927, Boyer [2] identified that there is considerable misunderstanding regarding corrosion of Mg, with the rapid corrosion of Mg attributed to the tendency of Mg itself to corrode. However, as described by Boyer, alloying elements (either deliberately added or impurities) have the most profound influence on corrosion of Mg. Since then, this statement has been supported by many demonstrations that alloying Mg can lead to wide variations in the corrosion rate [3]. However, from a practical perspective, pure Mg is incapable of providing the mechanical properties required for many implant applications [4, 5]. Therefore, potential alloying elements need to be carefully considered. Important properties of Mg include [6]: (1) hexagonal crystal structure, (2) an atomic diameter of 0.320 nm (meaning that there is a favourable size factor with elements such as Zn, Al, Ce, La, Ag, Zr) and (3) compared to other engineering metals, the number of elements with solid

N. T. Kirkland and N. Birbilis, *Magnesium Biomaterials*, SpringerBriefs in Materials, DOI: 10.1007/978-3-319-02123-2_4, © The Author(s) 2014

Table 4.1 Maximum solubility data (taken at the temperature where solubility is at a maximum) for binary Mg alloys from [7]

Element	at. %	wt%	System
Aluminium	11.8	12.7	Eutectic
Calcium	0.82	1.35	Eutectic
Cerium	0.1	0.5	Eutectic
Gold	0.1	0.8	Eutectic
Lithium	17	5.5	Eutectic
Manganese	1.0	2.2	Peritectic
Neodymium	1	3	Eutectic
Silver	3.8	15	Eutectic
Tin	3.35	14.5	Eutectic
Titanium	0.1	0.2	Peritectic
Yttrium	3.75	12.5	Eutectic
Zinc	2.4	6.2	Eutectic
Zirconium	1	3.8	Peritectic
Cadmium*	*100*	*100*	Solid solution
Indium*	*19.4*	*53.2*	Peritectic
Lead*	*7.75*	*41.9*	Eutectic
Scandium*	*15*	*24.5*	Peritectic
Thallium*	*15.4*	*60.5*	Eutectic

*Included to indicate systems of high solubility, however, they are not relevant on the basis of toxicity (i.e. lead, cadmium) and/or cost (i.e. scandium)

solubility in Mg, and the extent of any solid solubility, is relatively low. Table 4.1 presents binary solubility data for selected elements in Mg.

The modern portfolio of commercial Mg alloys has evolved to include compositions that containing aluminium (Al), zinc (Zn), calcium (Ca), rare earths (RE), lithium (Li), manganese (Mn) and zirconium (Zr). Such alloys have been developed on the basis of physical properties, such as strength, ductility, creep resistance or ease of production (i.e. castability). As such, the vast majority (very close to all) of biocorrosion studies to date have been performed on commercial alloys, which are not necessarily tailored to perform as biomaterials.

Of all the available elements, perhaps the most controversial is Al. Al is the most common alloying addition to structural Mg alloys [8], and adding Al increases mechanical properties and ease of castability (for example, a strength increase while retaining ductility is possible with AZ31, and appreciable strength is possible from high-pressure die cast AZ91 components). Many Mg alloy biocorrosion studies have been carried out on AZ (Al–Zn containing) Mg alloys. In such alloys, the proportion of Al can be high (9 %), and hence, the dose of Al from a dissolving implant must be factored into biocompatibility. Several studies have found few if any negative side effects when testing Al-containing Mg alloys both in vitro [9–11] and in vivo [12, 13]. However, these studies were typically short term and may have been heavily influenced by the corrosion of the alloy itself, especially in in vitro tests where platelet adhesion [11] or similar methods are used. In such cases, it is realistically impossible to isolate the effect an increased

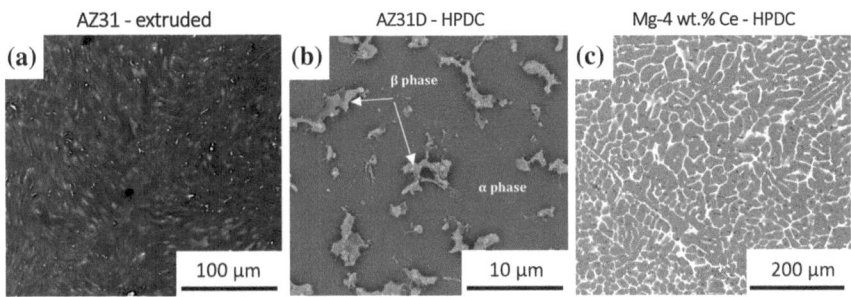

Fig. 4.1 Typical scanning electron micrographs for **a** extruded AZ31, **b** high-pressure die cast AZ91, **c** high-pressure die cast binary Mg-4 wt.% Ce

corrosion rate might have on the perceived toxicity of the investigated alloy. While alloy toxicity is described below, it is mentioned here to indicate that indiscriminate use of existing commercial alloys is not prudent. A more holistic understanding of Mg alloy biomaterials may well require a systematic study commencing from the binary systems and moving towards more complex systems.

In this chapter, we will not cover Mg alloy properties in detail; instead, readers are referred to dedicated works [8, 14] for discussion of physical properties such as strength. It is important to note here that Mg alloys will almost always possess relatively heterogeneous microstructures, because the alloying elements are sparingly soluble. For example, above ~ 3 wt.%, the room temperature solubility limit for Al is reached and the intermetallic phase $Mg_{17}Al_{12}$ (β-phase) is present [6]. Similarly for Mg–RE or Mg- x -RE alloys (where x represents an additional ternary element), intermetallics of the $(Mg, x)_x RE_y$ form are present [15]. Such intermetallics, which may be precipitates from careful heat treatment, or formed during solidification in die casting, contribute to strength. However, they also—as described below—increase localised corrosion of the Mg alloy. Images of typical microstructures of Mg alloys are shown in Fig. 4.1.

4.1.2 Corrosion of Mg Biomaterials

Alloying additions to Mg have a profound influence on the rate of corrosion. This has been understood prior to the large-scale production of magnesium in the 1930s, via the works from Boyer in the late 1920s [2], and then the seminal works of Hanawalt [16, 17] in the 1940s. Frankly speaking, only small advances in the metallurgical aspects of corrosion of Mg have occurred since then. The factors which control the corrosion of Mg remain those outlined by Boyer and Hanawalt:

1. Alloy purity—above so-called threshold limits corrosion rapidly accelerates. Mg alloys should only be used when the composition is below the "threshold limits" for Fe, Ni and Cu [17].

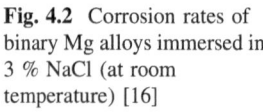

Fig. 4.2 Corrosion rates of
binary Mg alloys immersed in
3 % NaCl (at room
temperature) [16]

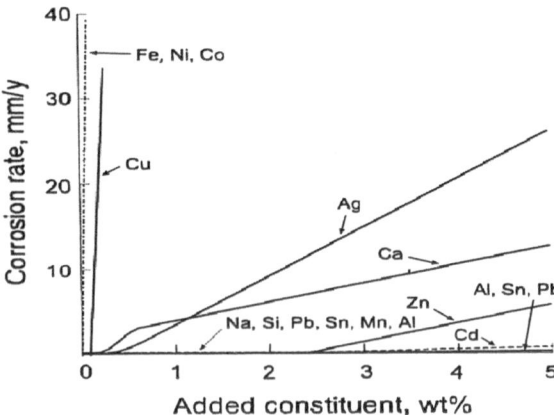

2. Scavenging of impurities—it was noted that the addition of Mn to Mg–Al
   alloys would scavenge any impurity Fe (by the formation of AlMnFe intermetallics) and hence significantly reduce the rate of corrosion [2].

Although biocorrosion is a process that occurs at physiological temperatures,
given the lack of comprehensive studies in the bio-Mg area, some results from
tests in saline solutions (NaCl) at room temperature will be included here. The
classical plot relating corrosion to alloying additions is given in Fig. 4.2.

It can be seen here that different elements clearly have significantly varied
effects on the resultant corrosion rate. As a consequence of this, Mg and its alloys
display corrosion rates that vary over several orders of magnitude, which is unique
for a metal system. A demonstration of this in the biocorrosion context is presented
in Fig. 4.3, where corrosion rates (following exposure to MEM for up to 14 days)
for a range of commercial and custom alloys are plotted in ascending order. The
corrosion rates cover over 3 orders of magnitude.

The spread of corrosion rates in Fig. 4.3 suggests that degradation can be
controlled within the 1–100 mm/yr range. Thus, the scope for "tunable" implants
is great, although there are complications, such as toxicity of alloying elements,
unsuitability (i.e. insolubility) of corrosion products, or heavily localised corrosion
(which may not be suitable for implants).

4.2 Crystalline Alloys

Table 4.1 listed the solubilities of relevant elements in crystalline Mg. What is not
appreciated from the table is that many elements are insoluble in Mg; this is
especially important for the design of bioresorbable alloys. Insolubility can be split
into three categories:

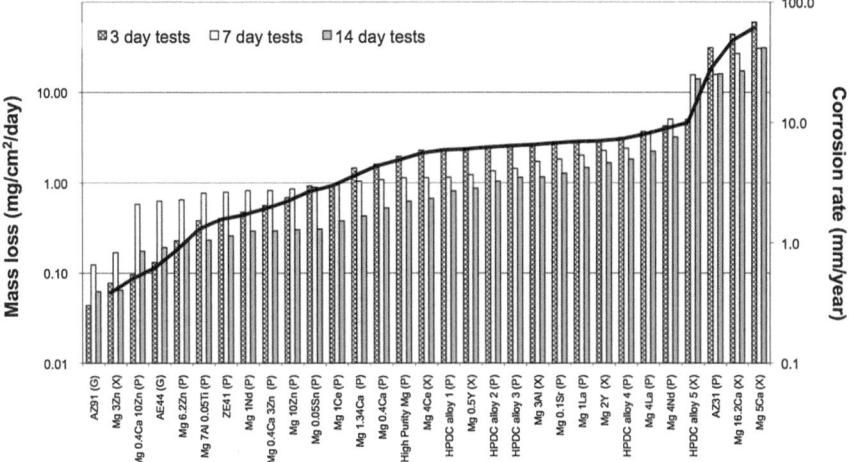

Fig. 4.3 Experimentally determined corrosion rates for Mg alloys tested at 37 °C in MEM. High purity magnesium was <40 ppm Fe. All compositions were given in weight percentage. The notation *G* refers to alloys which suffered a general corrosion mode, *P* refers to a pitting corrosion mode, and *X* refers to extremely localised corrosion [3]

1. *Complete and full insolubility*, where no solid solution is formed, and no Mg intermetallic forms with the particular alloying element. This situation is typical of Fe (and Mo, Nb) as shown in Fig. 4.4a. When Fe is combined with Mg, it exists in the form of pure Fe (bcc) particles, with no mutual solubility. This situation causes dramatic corrosion.

2. *No solubility,* but the formation of a Mg intermetallic, which is typically what is seen with Si (and Cu, Co, Ni). In such cases, an Mg_2X (X = Si, Cu, Co, Ni) intermetallic will form, because the elements have no room temperature solubility (Fig. 4.4b), and thus, the retention of a homogenous (single-phase) microstructure is not possible. Elements which form the Mg_2X intermetallic are often very problematic from a corrosion perspective, as they enhance the cathodic reaction because of the large exchange current density of metals such as Cu and Ni.

3. *Some solubility with no Mg intermetallic* is possible for elements such as Zr (Fig. 4.4c). In such cases, the alloying element will enter the solid solution to a limited extent, after which any further alloying will result in a separate, pure phase of the element. Again, such phases are problematic from a corrosion perspective, particularly at high concentrations.

4.2.1 Influence of Alloying Elements on Dissolution of Mg

A brief summary of the influences of common elemental additions to Mg is given here.

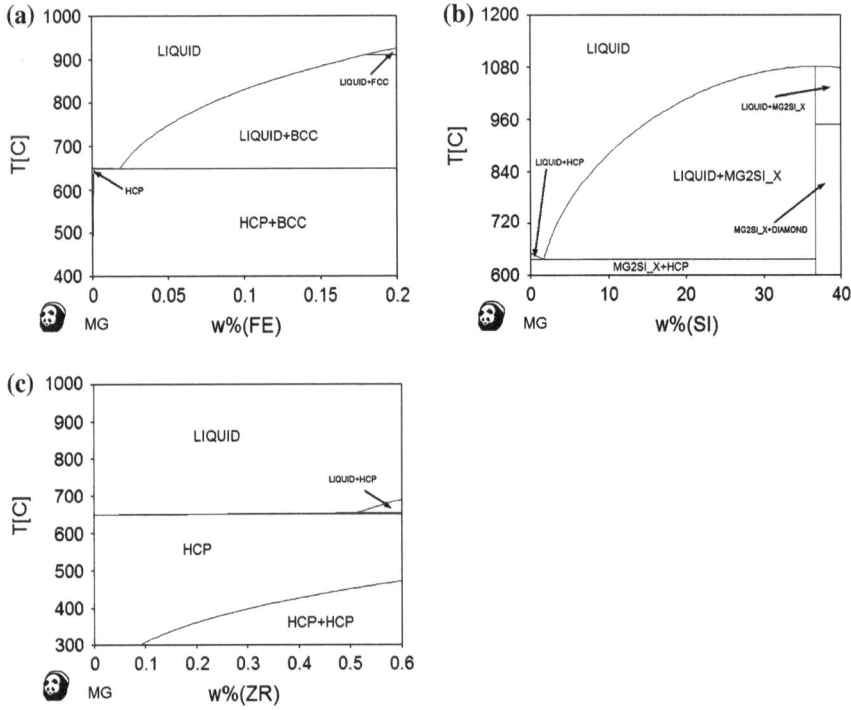

Fig. 4.4 Calculated binary phase diagrams for **a** Mg–Fe, **b** Mg–Si, **c** Mg–Zr

4.2.1.1 Aluminium

Al is the most common addition to Mg alloys. Al additions below the solubility limit tend to be inconsequential and can slightly decrease Mg corrosion by lowering the rate of the anodic reaction [18]. Because Al is more noble than Mg, Mg–Al alloys tend to have E_{corr} ennobled by ~ 100 mV compared to that of pure Mg. Mg–Al alloys, in particular AZ31, have the lowest corrosion rates of the commercial Mg alloys [8]. Above the solubility limit, excess Al additions tend to enhance the cathodic reaction; although this further ennobles E_{corr}, it also increases the corrosion rate as stimulated by the enhanced reduction reaction rates upon β-phase ($Mg_{17}Al_{12}$).

4.2.1.2 Calcium

In the binary context, Ca additions at levels below the solubility limit either do not affect or slightly increase corrosion rates in Mg, and additions above the solubility limit (which is ~ 1 wt.%) give rise to exceptionally high corrosion rates. Mg–Ca alloys have among the highest reported corrosion rates for candidate bioresorbable

Mg alloys and produce an insoluble, voluminous corrosion product [3]. Elemental Ca is even more reactive than Mg. As such, rather than the typical scenario of corrosion rate increasing by enhancement of cathode kinetics, Ca increases corrosion rates by monotonically enhancing anodic kinetics [19].

4.2.1.3 Iron

With the prevalence of steel tooling and production equipment, Fe is an unfortunately common, detrimental impurity in Mg alloys. Because of the very low solubility limit in Mg (~ 0.001 wt.%), Fe forms strong microgalvanic couples with Mg under open-circuit conditions (where Fe is polarised >1 V into its cathodic zone), hence serving as an intense local cathode and significantly increasing the corrosion rate of Mg. The literature reports a number of tolerance limits for Fe; typically, the Fe content must be less than 0.005 wt.% [8, 14].

4.2.1.4 Lithium

Mg–Li alloys exhibit filiform corrosion and have a non-compact surface oxide film [20]. Overall, Li is reported to increase the corrosion rate of Mg, and thus, Mg–Li alloys are deemed to have high corrosion rates [21]. Reported corrosion rates are not particularly consistent, and the reasons for this variation in corrosion behaviour still require clarification.

4.2.1.5 Manganese

Mn is typically added to Mg–Al and Mg–Al–Zn systems to reduce corrosion rate via the incorporation of essentially insoluble metals into an Al–Mn-X intermetallic phase; the classic example of this is the incorporation of Fe into Al–Mn–Fe [2, 16].

4.2.1.6 Rare Earths (Cerium, Lanthanum, Neodymium)

In binary alloys with Mg, Ce and La form $Mg_{12}Ce$ and $Mg_{12}La$, respectively, these intermetallics can accelerate corrosion by enhancing the cathodic kinetics [22]. In spite of the chemical reactivity of Ce and La, $Mg_{12}Ce$ and $Mg_{12}La$ are nobler than Mg and sustain the cathodic reaction at higher rates than Mg over the typical range of potentials of Mg alloys. The addition of Nd to Mg accelerates corrosion because of formation of the Mg_3Nd intermetallic, which also serves as a local cathode to the Mg matrix [22]. However, the increase in corrosion from Nd additions is less than that arising from Ce or La. As such, if addition of RE elements is desired, Nd is the best element to add; however, it is also the most expensive. In AZ and AM alloys, REs forms Al–RE intermetallics [23], which can inhibit precipitation and

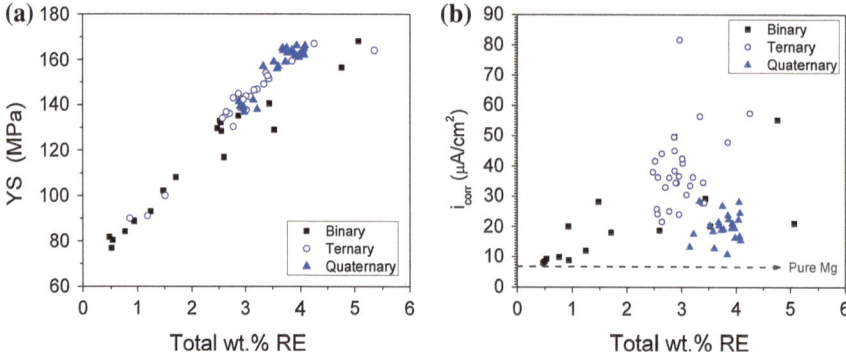

Fig. 4.5 a Yield strength versus total RE content arising from additions of a single RE to Mg (i.e. to give a binary alloy) and additions of multiple REs (to give ternary and quaternary alloys). **b** Corresponding corrosion rates [24]

reduce the volume fraction of $Mg_{17}Al_{12}$. From a corrosion perspective, however, Al–RE intermetallics are problematic because they act as cathodes.

The addition of RE elements to Mg represents an appropriate point to describe some of the Mg alloy design trade-offs and characteristics that should be considered when designing bioresorbable magnesium alloys. It is shown in Fig. 4.5a that irrespective of which RE element is used (including combinations of REs), the strength increases monotonically with total RE content. However, as shown in Fig. 4.5b, the influence of RE additions on corrosion rate depends upon the type and blending of the RE elements. In other words, the particular elements and their chemistry play a key role in determining the electrochemical response of Mg alloys, but a smaller role in dictating the mechanical properties. These data presented in Fig. 4.5 show that there is scope for customising the strength (by selecting the total RE content, Fig. 4.6a) and dissolution rates (by selecting the make-up of that RE content to exploit the vertical spread in Fig. 4.5b) of Mg alloys. The tunability of dissolution rates is demonstrated in Fig. 4.6, revealing that a very well controlled spread of corrosion rates can be achieved by composition control. However, we believe that the use of RE elements in a biodegradable context will require longer-term toxicity trials (both in vitro and in vivo) in coming years.

4.2.1.7 Yttrium

In a binary alloy with Mg, adding Y monotonically increases corrosion rate as the volume fraction of $Mg_{24}Y_5$ increases [25]. Other works have reported a minor decrease in corrosion when Y is added to Mg–Al alloys [26], and attributed this to refined microstructure, and a change in the morphology of the $Mg_{17}Al_{12}$ phase dispersion with an associated decrease in its volume fraction [27]. The corrosion rates of Mg–Y alloys in 0.1 M NaCl at room temperature are shown in Fig. 4.7 for

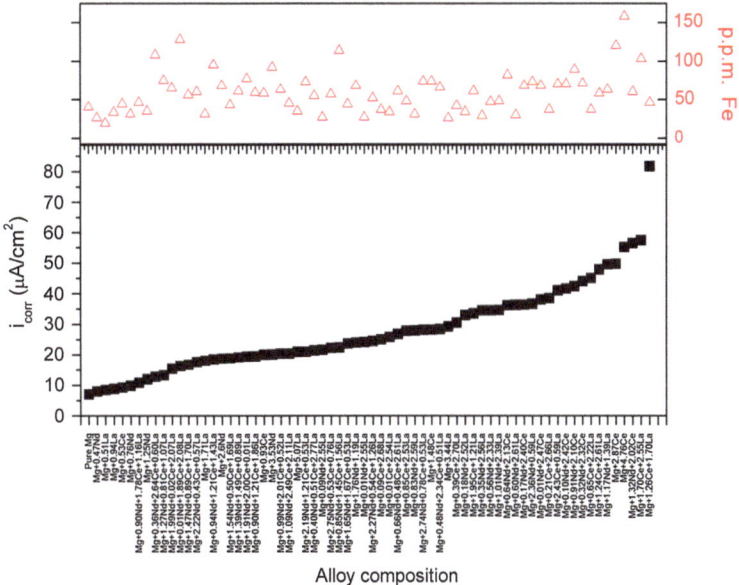

Fig. 4.6 Spread of corrosion rates sorted to reveal that composition can be exploited to control corrosion rates of Mg–RE alloys [24]

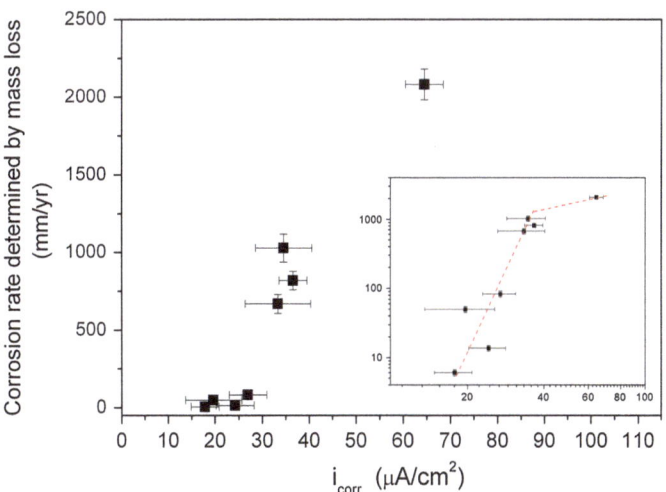

Fig. 4.7 Corrosion rate (in mm/yr) determined from mass loss testing, compared with electrochemically determined values of i_{corr} for binary Mg–Y alloys with 0–18 wt.% Y (Inset is same data represented using a *log-scale*) [28]

additions of Y from 0 to 18 wt.%. At high Y loadings, corrosion rates increase excessively; this is because the relationship between corrosion rate and %Y is logarithmic.

4.2.1.8 Zinc

Zinc is often added to Mg–Al and Mg–Zr alloys to provide solution strengthening. Binary (Mg–Zn) alloys were also explored as potential biodegradable metals [29]. Adding up to 20 %, Zn monotonically increased the rate of corrosion at physiological temperature in MEM.

4.2.1.9 Zirconium

Corrosion performance of Zr-containing alloys, such as the ZK and ZE series, is relatively satisfactory, although not as good as that of the AZ class of alloys. It has been reported that corrosion resistance may be improved by heat treatment to be comparable to that of Al-based Mg alloys [6]. There are also reports that zirconium contents of up to ~ 0.42 wt.% have minimal influence on corrosion [30]; however, excessive additions of Zr will lead to insoluble Zr and formation of pure Zr particles in the matrix, which are detrimental to corrosion [31]. Zr is not generally added to Mg–Al alloys, as it combines with Al to form an Al_3Zr intermetallic that negates the grain refining effect of Zr. Until rather recently, the influence of Zr on Mg had not received specific attention, because this required the production of alloys with varying Zr content to isolate the effect of Zr. Results to date suggest that Zr at low levels in multielement alloys is innocuous; however, excess Zr will cause significant corrosion, and Mg–Zr binary alloys also suffer excessive corrosion [32].

4.3 Amorphous Alloys

4.3.1 Mg–Glass Biomaterials

Metallic glasses, or amorphous alloys, are non-crystalline solids exhibiting the phenomenon of a glass transition [33]. Amorphous alloys were first discovered some 50 years ago [34]. Such alloys are produced by rapid cooling from the melt. In practice, to achieve the required large rate of heat extraction, at least one dimension of the resulting solid must be small (e.g. small thickness). As such, a critical casting diameter is used to indicate glass-forming ability (GFA) of amorphous alloys. The first amorphous alloys, and the majority studied to date, are ribbons or wires, several microns thick, produced by melt spinning. These thin ribbons or wires are unsuitable for biomaterial applications, because they cannot bear sufficient load. However, following on from the 1974 discovery of a "bulk" metallic glass (BMG) [35], it is also possible for small bulk portions of glass to be achieved. Please note that here the word "bulk" refers to amorphous metals with a thickness greater than one millimetre. As such, BMGs are still quite thin, typically no more than 2 mm thick.

All BMGs consist of at least three elements. Most BMG development has, including in recent years, been empirical; however, BMGs tend to have a significant difference in atomic size, with size ratios >12 % among the three main constituent elements. The main constituent elements also should have negative heats of mixing. Amorphous Mg–Zn ribbons were first reported in 1977 [36]. However, the most relevant glasses for biodegradable applications are BMGs, based on the Mg–Zn–Ca ternary alloy system; these are an emerging class of material that offer properties quite different from those of conventional crystalline magnesium alloys [37, 38].

4.3.2 Corrosion of Mg–Glass Biomaterials

The study of corrosion of amorphous metals to date has shown that generally amorphous alloys have superior corrosion characteristics compared to those of their crystalline counterparts [39–45]. This has been attributed to: (1) chemical homogeneity (2) monolithic structure and (3) lack of grain boundaries. In particular, studies on Ni- [40] and Zr-based [41, 42] BMGs have concluded that the amorphous alloys corrode less than their crystalline counterparts. However, in our appraisal of this field we found that many tests were conducted using quite mild environments; for example, selection of a test pH where the alloy is expected to remain passive. There is far less conclusive information available regarding the corrosion of BMGs under depassivating conditions. In the case of Mg-based BMGs, physiological exposure and neutral pH would result in an environment where the BMG is not spontaneously passive.

The most relevant system for bioresorbable metallic glasses is the Mg–Ca–Zn system, defined by the phase diagram in Fig. 4.8.

Of the reports to date regarding corrosion of Mg–Ca–Zn BMGs, Wang et al. [45] studied the corrosion of amorphous $Mg_{67}Zn_{28}Ca_5$ ribbon and its crystalline counterpart, reporting that the amorphous ribbon exhibited a more noble corrosion potential, as well as a lower corrosion current density. In a separate study, Wang et al. [43] evaluated the corrosion of $Mg_{67}Zn_{28}Ca_5$, from the amorphous to crystalline states, including a range of semi-crystalline alloys with various degrees of crystallinity. A combination of hydrogen evolution and potentiodynamic polarisation tests showed that the as-quenched amorphous ribbon corroded more than a semi-crystalline sample (prepared by annealing the amorphous ribbon at 160 °C). The study did not compare the corrosion behaviour of these alloys to that of more conventional crystalline Mg alloys containing smaller amounts of alloying elements.

Cytocompatibility tests on $Mg_{66}Zn_{30}Ca_4$ and $Mg_{70}Zn_{25}Ca_5$ BMGs, reported by Gu et al. [47], revealed that cell viability was higher on the BMGs than on

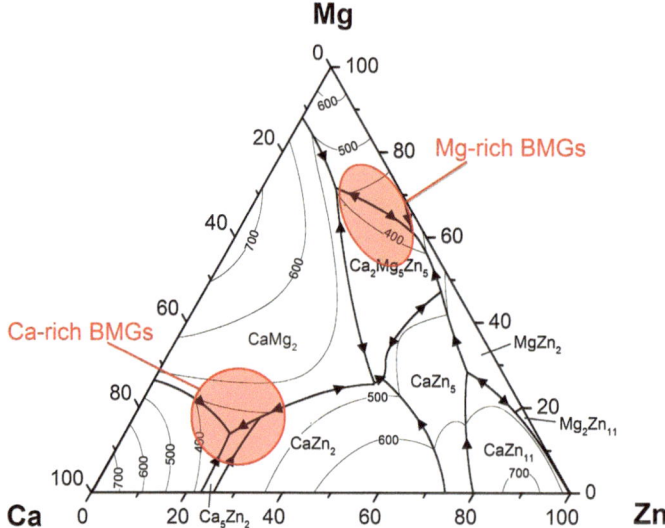

Fig. 4.8 Ternary phase diagram of the Mg–Ca–Zn system, showing the compositional regions where bulk metallic glasses may form [46]

as-rolled pure Mg. L929 and MG63 cells adhered well and proliferated on the surface of the $Mg_{66}Zn_{30}Ca_4$ samples. In a separate study by Zberg, a reduction in hydrogen evolution associated with the corrosion of Zn-rich Mg–Zn–Ca glasses was reported [37]. Zberg suggested that above a Zn-alloying threshold of ∼28 at.%, Zn- and oxygen-rich passivating layers form on the alloy surface in simulated body fluid. However, in their work, the detailed testing was executed at room temperature, where the dissolution rates measured are significantly lower than those at physiological temperature. Similar alloy compositions were tested by Cao at physiological temperatures, and the in vitro corrosion rates were much higher [38]. The study from Zberg (and indeed most other studies of bio-Mg BMG) states enthusiastically that glassy Mg–Ca–Zn alloys show great potential as biomaterials. However, one key obstacle to be overcome is that the critical casting thickness for such BMGs is still very small (compared to the desired size of many implants) at present. To some extent, forming in the glassy state allows thin, stiff sections to be produced, which will be appropriate for some applications. Larger specimens will be affected by crystallisation, which increases corrosion [48].

A comprehensive report by Cao [49], which paid particular attention to the critical casting thickness, found that within the family of Mg–Zn–Ca BMGs, Ca-rich compositions (i.e. Ca-based BMGs) yield larger critical casting thicknesses and appreciable strengths. In the context of biodegradable implants, toxicity of Ca is not a concern, because the daily allowable Ca dose for humans is well over 1,000 mg [50].

The in vitro corrosion performance of a range of BMGs was measured at 37 °C in MEM, and the dependence of corrosion behaviour on composition was very

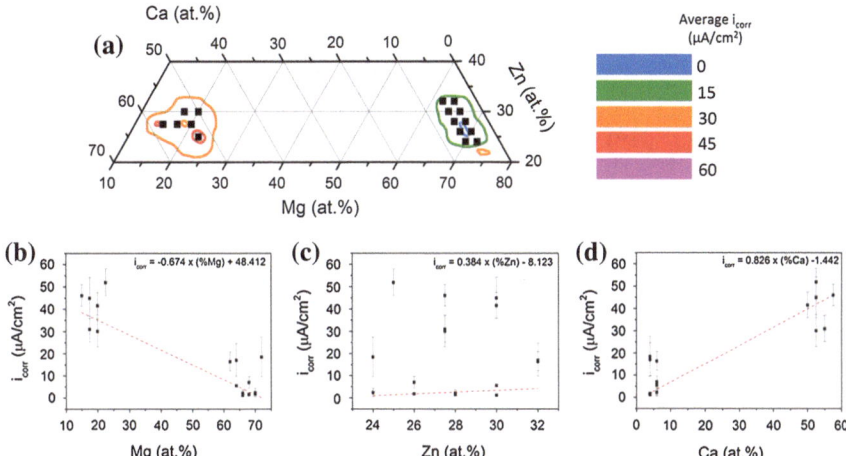

Fig. 4.9 a Dissolution rate displayed as a contour plot on the Mg–Zn–Ca ternary diagram. The sensitivity of corrosion rate to **b** Mg, **c** Zn and **d** Ca; *dashed lines* are regression fits [38]

complex (Fig. 4.9). A monotonic increase in Zn content in both Ca-rich and Mg-rich BMGs did not reduce their corrosion. This suggests that the corrosion rate is not solely determined by the Zn content but, rather, the Mg:Zn:Ca ratio. What is also obvious is that because Ca is more reactive, Ca-based BMGs have significantly higher biocorrosion rates than those of Mg-based BMGs. Cao showed in a further study that rates of dissolution for Ca-based BMGs were much higher than those of conventional Mg alloys, meaning that Ca-based BMGs were only suited to applications requiring short-term implants [51]. In a study by Wang [52], Ca–Mg–Zn BMGs were tested; although the comment was made that these glasses may be suitable biomaterials, the Ca–Mg–Zn BMG was very rapidly biodegradable. The degradation time in HBSS was as short as three hours, and little evidence of any BMG was found four weeks after implantation in a mouse marrow cavity. Nonetheless, Ca–Mg–Zn BMG extracts showed good biocompatibility over a wide range of concentrations and did not cause death of L929, VSMC or ECV304 cells, as might have been expected Table 4.2.

4.4 Alloy Design Considerations

4.4.1 Alloying Mg for Biomaterials—Toxicity Considerations

Although not the focus of this monograph, the short- and long-term effects of introducing foreign metals (most likely as ions) into the human body present an important alloy design consideration. The alloy designer nominally has the

Table 4.2 Summary of i_{corr} and E_{corr} of various BMGs (error refers to the standard deviation in each data set) [38]

Alloy name	Mg (at. %)	Zn (at. %)	Ca (at. %)	Avg i_{corr} ($\mu A/cm^2$)	i_{corr} error ($\mu A/cm^2$)	Avg E_{corr} (mV_{SCE})	E_{corr} error (mV_{SCE})
MgZnCa1	72	24	4	18.4	8.9	−1321	14
MgZnCa2	70	26	4	1.7	0.2	−1238	31
MgZnCa3	68	28	4	1.5	0.7	−1265	14
MgZnCa4	66	30	4	1.1	0.4	−1264	10
MgZnCa5	64	32	4	17.0	7.4	−1196	22
MgZnCa6	70	24	6	2.3	1.9	−1242	27
MgZnCa7	68	26	6	6.9	2.6	−1192	26
MgZnCa8	66	28	6	2.2	1.4	−1194	22
MgZnCa9	64	30	6	5.5	0.5	−1207	60
MgZnCa10	62	32	6	16.4	4.3	−1198	11
CaMgZn11	15	27.5	57.5	46.0	4.8	−1435	6
CaMgZn12	17.5	27.5	55	30.9	5.8	−1398	23
CaMgZn13	20	27.5	52.5	30.0	7.2	−1398	12
CaMgZn14	17.5	30	52.5	44.8	9.2	−1411	12
CaMgZn15	22.5	25	52.5	51.8	6.0	−1419	9
CaMgZn16	20	30	50	41.5	5.8	−1411	16

complete periodic table within their grasp (with some obvious exceptions); however, many inorganic substances are toxic to the human body. For example, long-term effects of exposure to Al are unclear. Animal studies have found excessive exposure to Al toxicity to result in a variety of problems, including affecting reproduction [53], inducing dementia [54] and potentially leading to Alzheimer's disease (although data are not conclusive for humans) [55, 56]. Similar biocompatibility concerns exist for alloys containing RE; although effects of RE have been studied in vitro [57–60] and in vivo [61, 62], there is a lack of knowledge of their long-term effects when implanted [51, 63]. This creates the potential that the significant body of work that has been and will be performed using alloys containing Al and/or RE may, in the end, be unexploited if these materials cannot be proven to be non-toxic. To this effect, the recommended daily dosage limits for RE elements (i.e. Ce, La, Nd, Pr, Y, Gd) need further study. Similarly, although Li has been used in medicine for almost 150 years [64, 65], it has not been employed widely in implanted materials—where continual exposure may occur on the mg/day level [66]. A compilation of approximate maximum daily dosage limits (retrieved from the literature) is presented in Table 4.3.

There are many and varied instances in the literature where the toxicity has not necessarily been considered during alloy design. An example where the long-term toxicity is unknown is a Sr-based BMG, $Sr_{40}Mg_{20}Zn_{15}Yb_{20}Cu_5$, that has been suggested as a potential biodegradable metal [67]. While initial cytotoxicity studies may reveal low toxicity, there is very little knowledge regarding the effect of potential overdoses of elements such as Sr or Yb, in particular, if these elements do any longer-term damage to organs or mental health. Similarly, a $Mg_{75}Cu_{13.33}Y_{6.67}Zn_5$ BMG

Table 4.3 Toxicity limits for Mg and typical alloying elements as reported in [29]

Element	Daily allowable dosage (mg)
Al	14
Be	0.01
Ca	1400
Ce	4.2
Cu	6
Fe	40
La	4.2*
Mg	400
Nd	4.2*
Ni	0.6
Pr	4.2*
RE	4.2*
Sn	3.5
Sr	5
Ti	0.8
Y	0.016*
Zn	15

*Denotes that the total amount of these rare earth elements (Ce, La, Nd, Pr, Y) should not exceed 4.2 mg/day

composite has been reported. This composite is formed in situ and has a plastic strain of about 19 % and compressive strength of about 1,040 MPa [68]. However, if this material is used as a biomaterial, high concentrations of Cu and Y may potentially be released into the body; the longer-term effects of this exposure are again not well known.

4.4.1.1 Corrosion Products

Another important factor in the biocorrosion of Mg is that corrosion products will need to be accommodated (by some means) in the body. This is particularly important if the ions released during corrosion form products (oxides, hydroxides, carbonates) rapidly, rather than migrating from the implant surface. This aspect of each potential biomaterial must be studied on a case-by-case basis; however, some examples are given below (Fig. 4.10).

In addition to the issue of corrosion products, localised corrosion must also be considered. Potentially, the issue of localisation of corrosion may be less of a concern for BMGs than for polycrystalline alloys; however, this is an area that has not received the attention it deserves. Cao suggested that corrosion of Ca-based BMGs occurs by a dealloying process [51]; this is probably also the case for longer-term dissolution of Mg-based BMGs.

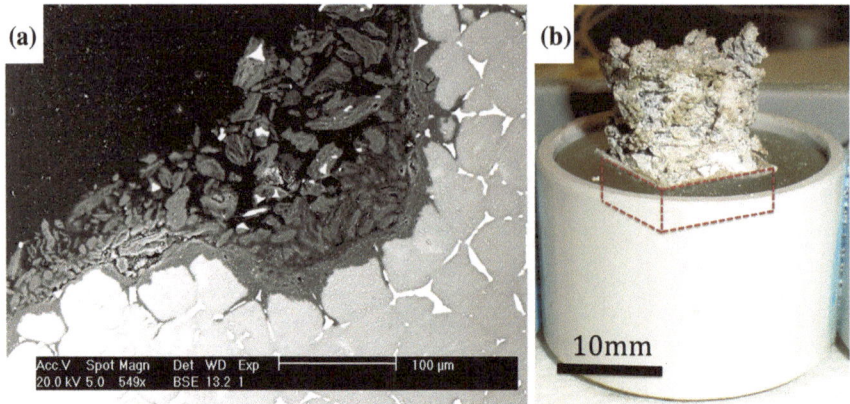

Fig. 4.10 **a** Corrosion products from Mg alloy ZE41 exposed to MEM at 37 °C. We observe that corrosion is not uniform **b** Excessively voluminous and insoluble corrosion products that form upon Mg–Ca (where Ca > ~5 %) during exposure to MEM at 37 °C [3]

4.4.2 A Case Study in Mg Biomaterial Alloy Design

The development of suitable biodegradable implant alloys is a multidisciplinary challenge, since the alloy design must be confined to biologically non-toxic alloying additions while still providing the requisite mechanical properties—and the required (often application-specific) dissolution rate. This leaves a small number of compatible elements that can deliver mechanical or corrosion benefits when alloyed with Mg. Ca and Zn are probably the most biocompatible [29]; however, corrosion acceleration from Ca, and subsequently, insoluble its oxide products remains a concern at high Ca loadings. An example of the performance-driven design of bioresorbable Mg was recently presented; in this study, binary Mg–Zn, binary Mg–Ca and ternary Mg–Ca–Zn/Mg–Zn–Ca alloys had their mechanical and biocorrosion performance mapped [29]. The results are presented in Fig. 4.11. With increased alloying, the microstructures become more heterogeneous (Fig. 4.111C, D). Ca is found to have a more potent strengthening effect than that of Zn (Fig. 4.11 [1]). Ca also produces a greater increase in the biocorrosion rate (Fig. 4.11 [2]). This is because Ca increases the rate of anodic dissolution, while the somewhat more subtle increase in the dissolution rates when Zn is added are because of enhancement of the cathodic reaction. As a consequence, these two elements have different strengthening propensities and different modes of increasing the biocorrosion rate. The result of this is that careful selection of composition by balancing the effects of both elements can increase strength and tune the biocorrosion rate [29]. Such a property-directed approach to alloy design is perhaps unique to the realm of biomaterials and bioresorbable implants. A neural network model for user-interactive alloy design based on this case study is available for use at (http://www.extras.springer.com).

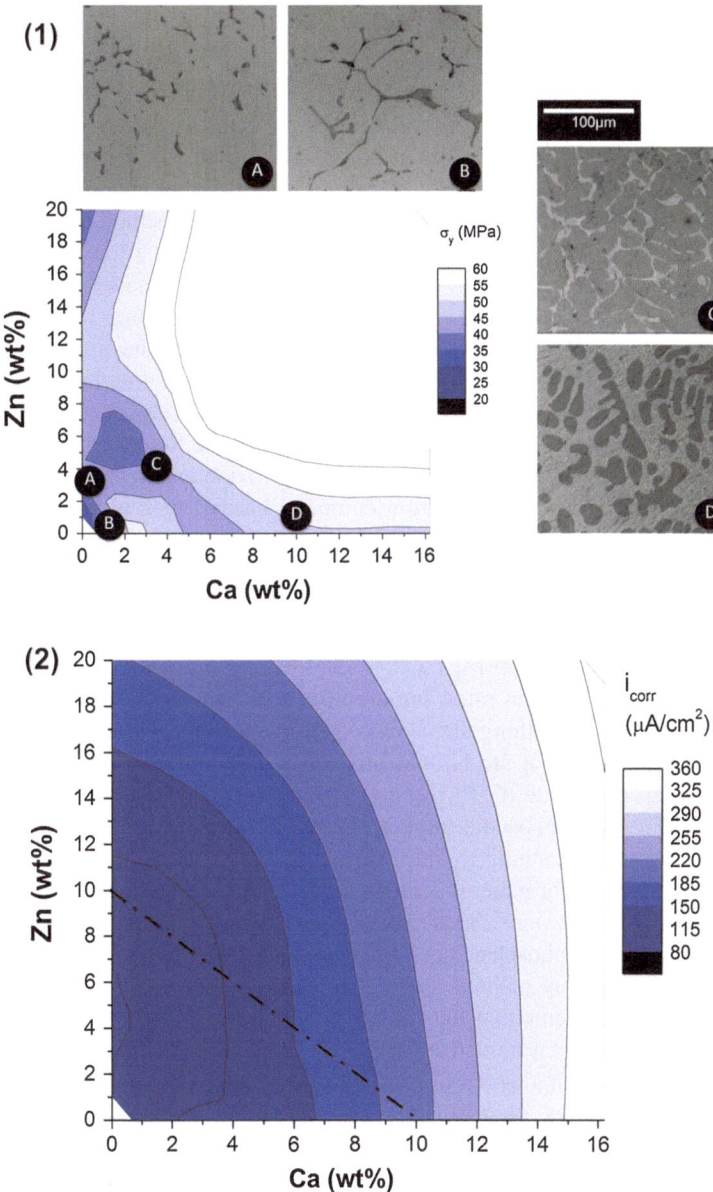

Fig. 4.11 (*1*) Yield strength of Mg alloys before immersion with micrographs of selected areas, (*2*) i$_{corr}$ of Mg alloys in MEM$_{FBS}$ [29]

4.4.3 Coating Mg and Biomimetic Coatings

Although the topic of coating magnesium cannot be fully covered here (for more detail readers are referred to [69]), some key aspects of relevance to design of bioresorbable materials will be discussed. Because pure Mg essentially displays the lowest biocorrosion rate of all crystalline Mg alloys, any situation requiring even slower corrosion rates will require the use of another method to achieve this. Also, in situations where the implant is not required to last a long period, rapid dissolution rates can occur; this means that some way of moderating the early dissolution (and associated hydrogen evolution) may be desirable. This is because, in the early stages of implant insertion, the acceptance of the implant, and the attachment of cells to the implant, could be frustrated by rapid initial corrosion. In addition to moderating corrosion during the first days of implantation, a coating might promote cell growth or attachment or stimulate some other favourable response. However, obviously, any coating must provide only a temporary level of protection, since if the coating were a complete barrier, then ultimate biodegradation could not occur, defeating the purpose of using resorbable implants. Coating of Mg alloys is thus another aspect of the implant that could be used to tune the level of corrosion protection and the longevity of the afforded protection.

As for the alloys, any coating must satisfy strict biocompatibility criteria. Many industrial coating systems for Mg (such as chromate conversion coatings) typically produce very low corrosion rates, but are also very toxic [70]. The ideal coating should therefore be something that slows corrosion rate in physiological solutions, contains no elements that are biologically unsafe and enhances the biocompatibility. Calcium phosphate (CaP) coatings are already well known to improve the biocompatibility of orthopaedic implants [71]. These coatings have been used on stainless steel and titanium to improve the biological response. As calcium phosphates are the major mineral component of bone, the coating is highly suitable for orthopaedic applications. Such coatings seem to be ideal for a biodegradable implant. The calcium phosphate coating can, in theory, be optimised to protect the Mg [72, 73], potentially without leaving any toxic by-products.

The Mg alloy systems that have been investigated as potential substrates for coated biomaterials are summarised in Fig. 4.12. It should be noted that coating these alloys is not a single-step process and that optimising the coating involves both carefully selected pretreatments and choice of coating composition. In addition, electroplated coatings require special attention [74].

While there are several studies of the performance of CaP coatings upon Mg, the example shown below has been taken from the work of Chen [75] (Fig. 4.13). In that work, a CaP coating was developed which included crystalline hydroxyapatite (HA), the most stable of the CaP compounds under physiological conditions [76]. The effectiveness of the coating at reducing biocorrosion rate was found to vary with the substrate alloy composition. Overall, there was a decrease in the mass lost for all alloys examined, except for Mg-3Zn. The tests were conducted for 24 h on alloys with relatively rapid dissolution rates (the Mg–Ca family), and in

Fig. 4.12 Mg alloys used as substrates in the evaluation of coatings on potential biomaterials [73]

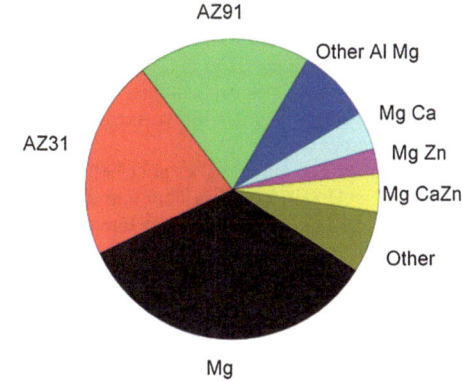

Fig. 4.13 Measured mass loss rates after exposure of various Mg alloy specimens to MEM at 37 °C for 24 h [75]

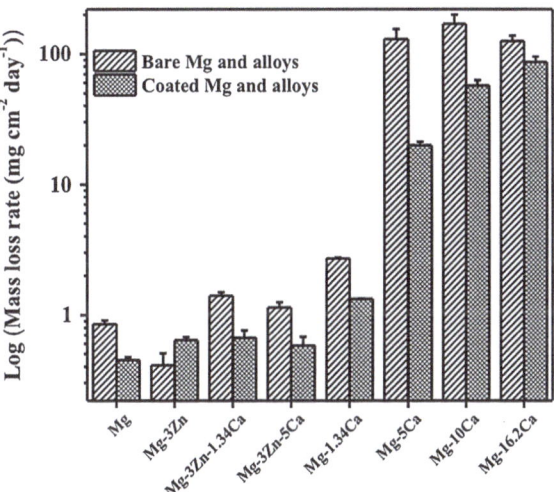

most cases, the early corrosion rate decreased by about half; this also corresponds to halving in the attendant hydrogen evolution. The relationship between alloy composition and effect of coating has not yet been studied in detail, but will need to be a focus of future investigations.

References

1. Survey USG (2011) Minerals Commodity Summary: Magnesium
2. Boyer JA (1927) The corrosion of magnesium and of the magnesium alloys containing manganese
3. Kirkland NT, Lespagnol J et al (2010) A survey of bio-corrosion rates of magnesium alloys. Corros Sci 52(2):287–291

4. Vojtěch D, Čížová H et al (2006) Investigation of magnesium-based alloys for biomedical applications. Kovove Mater 44:211–223

5. Ganeshan S, Shang SL et al (2009) Effect of alloying elements on the elastic properties of mg from first-principles calculations. Acta Mater 57(13):3876–3884

6. Polmear IJ (1989) Light alloys: metallurgy of the light metals, 2nd edn. Hodder & Stoughton, Melbourne

7. Massalski TB, Hiroaki O et al (1990) Binary alloy phase diagrams. ASM International, OH

8. Polmear IJ (1999) Magnesium and magnesium alloys. In: Avedesian MM, Baker H (eds) ASM specialty handbook. Asm International, USA, pp 12–25

9. Witte F, Feyerabend F et al (2007) Biodegradable magnesium-hydroxyapatite metal matrix composites. Biomaterials 28(13):2163–2174

10. Witte F, Feyerabend F et al (2006) Unphysiologically high magnesium concentrations support chondrocyte proliferation and redifferentiation. Tissue Eng 12(12):3545–3556

11. Gu X, Zheng Y et al (2009) In vitro corrosion and biocompatibility of binary magnesium alloys. Biomaterials 30(4):484–498

12. Duygulu O, Kaya RA et al (2007) Investigation on the potential of magnesium alloy AZ31 as a bone implant. Mater Sci Forum 546–549:421–424

13. Witte F, Abeln I et al (2008) Evaluation of the skin sensitizing potential of biodegradable magnesium alloys. J Biomed Mater Res, Part A 86A(4):1041–1047

14. Friedrich HE (2006) Magnesium technology: metallurgy, design data, Applications. Springer, Heidelberg

15. Chia TL, Easton MA et al (2009) The effect of alloy composition on the microstructure and tensile properties of binary mg-rare earth alloys. Intermetallics 17(7):481–490

16. Hanawalt JD, Nelson CE et al (1942) Corrosion studies of magnesium and its alloys. Trans AIME 147:273–299

17. McNulty RE, Hanawalt JD (1942) Some corrosion characteristics of high purity magnesium alloys. J Electrochem Soc 81(1):423

18. Makar GL, Kruger J (1993) Corrosion of magnesium. Int Mater Rev 38(3):138–153

19. Kirkland NT, Birbilis N et al (2010) In-vitro dissolution of magnesium–calcium binary alloys: clarifying the unique role of calcium additions in bioresorbable magnesium implant alloys. J Biomed Mater Res B Appl Biomater 95B(1):91–100

20. Song Y, Shan D et al (2009) Biodegradable behaviors of AZ31 magnesium alloy in simulated body fluid. Mater Sci Eng, C 29(3):1039–1045

21. Zhang C, Huang X et al (2008) Electrochemical characterization of the corrosion of a Mg-Li alloy. Mater Lett 62(14):2181–2184

22. Birbilis N, Easton MA et al (2009) On the corrosion of binary magnesium-rare earth alloys. Corros Sci 51(3):683–689

23. Arrabal R, Matykina E et al (2012) Corrosion behaviour of AZ91D and AM50 magnesium alloys with Nd and Gd additions in humid environments. Corros Sci 55:351–362

24. Birbilis N, Cavanaugh MK et al (2011) A combined neural network and mechanistic approach for the prediction of corrosion rate and yield strength of magnesium-rare earth alloys. Corros Sci 53(1):168–176

25. Sudholz AD, Kirkland NT et al (2011) Electrochemical properties of intermetallic phases and common impurity elements in magnesium alloys. Electrochem Solid-State Lett 14(2):C5–C7

26. Zhang J, Niu X et al (2009) Effect of Yttrium-Rich Misch metal on the microstructures, mechanical properties and corrosion behavior of die cast AZ91 alloy. J Alloy Compd 471(1–2):322–330

27. Li X, Jiao F et al (2012) Influence of second-phase precipitates on the texture evolution of Mg–Al–Zn alloys during hot deformation. Scripta Mater 66(3–4):159–162

28. Sudholz AD, Gusieva K et al (2011) Electrochemical behaviour and corrosion of Mg-Y alloys. Corros Sci 53(6):2277–2282

29. Kirkland NT, Staiger MP et al (2011) Performance-driven design of biocompatible Mg-alloys. JOM 63(6):28–34

30. Sun M, Wu G et al (2009) Effect of Zr on the microstructure, mechanical properties and corrosion resistance of Mg–10Gd–3Y magnesium alloy. Mater Sci Eng, A 523(1–2):145–151
31. Gandel D, Birbilis N et al (2010) Influence of manganese, zirconium and iron on the corrosion of magnesium. In: Proceedings of corrosion & prevention 2010, Adelaide, SA, Australia, 14 Nov 2010. Australian Corrosion Association, pp 875–885
32. Gandel DS, Easton MA et al (2013) Influence of Mn and Zr on the corrosion of Al-Free Mg-alloys, part ii: impact of Mn and Zr on Mg-Alloy electrochemistry and corrosion. Corrosion
33. Zarzycki J, Scott WD et al (1991) Glasses and the vitreous state. Cambridge University Press, Cambridge
34. Klement W, Willens RH et al (1960) Non-crystalline structure in solidified gold-silicon alloys. Nature 187:869–870
35. Chen HS (1974) Thermodynamic considerations on the formation and stability of metallic glasses. Acta Metall Mater 22(12):1505–1511
36. Calka A, Madhava M et al (1977) A transition-metal-free amorphous alloy: Mg70Zn30. Scr Metall 11(1):65–70
37. Zberg B, Uggowitzer PJ et al (2009) MgZnCa glasses without clinically observable hydrogen evolution for biodegradable implants. Nat Mater 8(11):887–891
38. Cao JD, Laws KJ et al (2012) Potentiodynamic polarisation study of bulk metallic glasses based on the Mg-Zn-Ca ternary system. Corros Eng, Sci Technol 47(5):329–334
39. Gu X, Zheng Y et al (2010) Corrosion of, and cellular responses to Mg-Zn-Ca bulk metallic glasses. Biomaterials 31(6):1093–1103
40. Liu S, Huang L et al (2012) Effects of crystallization on corrosion behaviours of a Ni-based bulk metallic glass. Int J Miner Metall Mater 19(2):146–150
41. Li WH, Chan KC et al (2012) Thermodynamic, corrosion and mechanical properties of Zr-based bulk metallic glasses in relation to heterogeneous structures. Mater Sci Eng A Struct Mater 534:157–162
42. Peter WH, Buchanan RA et al (2002) Localized corrosion behavior of a zirconium-based bulk metallic glass relative to its crystalline state. Intermetallics 10(11):1157–1162
43. Wang Y, Tan MJ et al (2012) In vitro corrosion behaviors of Mg67Zn28Ca5 alloy: from amorphous to crystalline. Mater Chem Phys 134(2–3):1079–1087
44. Andreas K, Jörg L et al (2007) Rapid solidification and bulk metallic glasses—processing and properties. In: Materials processing handbook. CRC Press, Boca Raton, pp 17–44
45. Wang Y, Tan MJ et al (2012) Corrosion performance of melt-spun Mg67Zn28Ca5 metallic glass in artificial sweat. J Mater Sci 47(18):6586–6592
46. Villars P, Prince A et al (1994) Handbook of ternary alloy phase diagrams. ASM International, Materials Park
47. Gu X, Zhou W et al (2010) Microstructure, mechanical property, bio-corrosion and cytotoxicity evaluations of Mg/Ha composites. Mater Sci Eng, C 30(6):827–832
48. Zhou X, Ralston KD et al (2013) Effect of the degree of crystallinity on the electrochemical behaviour of Mg65Cu25Y10 and Mg70Zn25Ca5 bulk metallic glasses. Corrosion
49. Cao JD (2013) Processing and properties of biocompatible metallic glasses. Doctor of Philosophy, The University of New South Wales, Sydney, Australia
50. Committee to review dietary reference intakes for Vitamin D and Calcium FaNB dietary reference intakes for calcium and vitamin D. In: Medicine Io (ed), Washington, DC, USA, 2010. National Academy Press
51. Cao JD, Kirkland NT et al (2012) Ca–Mg–Zn bulk metallic glasses as bioresorbable metals. Acta Biomater 8(6):2375–2383
52. Wang YB, Xie XH et al (2011) Biodegradable CaMgZn bulk metallic glass for potential skeletal application. Acta Biomater 7(8):3196–3208
53. Domingo JL (1995) Reproductive and developmental toxicity of aluminum: a review. Neurotoxicol Teratol 17(4):515–521
54. Lucey TD, Venugopal B (1977) Metal toxicity in mammals. Plenum Press, New York
55. El-Rahman SSA (2003) Neuropathology of aluminum toxicity in rats (Glutamate and GABA Impairment). Pharmacol Res 47(3):189–194

56. Flatten TP (2001) Aluminium as a risk factor in Alzheimer's disease, with emphasis on drinking water. Brain Res Bull 55(2):187–196
57. Feyerabend F, Fischer J et al (2010) Evaluation of short-term effects of rare earth and other elements used in magnesium alloys on primary cells and cell lines. Acta Biomater 6(5):1834–1842
58. Gruhl S, Witte F et al (2009) Determination of concentration gradients in bone tissue generated by a biologically degradable magnesium implant. J Anal At Spectrom 24(2):181–188
59. Bondemark L, Kurol J et al (1994) Orthodontic rare earth magnets—in vitro cytotoxicity assessment. Br J Orthod 21:335–341
60. Donohue VE, McDonald F et al (1995) In vitro cytotoxicity testing of neodymium-iron-boron magnets. J Appl Biomater 6(1):69–74
61. Zhang H, Zhu WF et al (1999) Subchronic toxicity of rare earth elements and estimated daily intake allowance. In: Ninth Annual V.M. Goldschmidt Conference, Cambridge, MA, USA
62. Bruce DW, Hietbrink BE et al (1963) The acute mammalian toxicity of rare earth nitrates and oxides. Toxicol Appl Pharmacol 5(6):750–759
63. Drynda A, Deinet N et al (2009) Rare earth metals used in biodegradable magnesium-based stents do not interfere with proliferation of smooth muscle cells but do induce the upregulation of inflammatory genes. J Biomed Mater Res, Part A 91A(2):360–369
64. Timmer RT, Sands JM (1999) Lithium Intoxication. J Am Soc Nephrol 10(3):666–674
65. Bhagwagar Z, Goodwin GM (2002) The role of lithium in the treatment of bipolar depression. Clin Neurosci Res 2(3–4):222–227
66. Yfantis CD, Yfantis DK et al (2006) New magnesium alloys for bone tissue engineering: in vitro corrosion testing. WSEAS Trans Environ Dev 2(8):1110–1115
67. Li HF, Zhao K et al (2012) Study on bio-corrosion and cytotoxicity of a Sr-Based bulk metallic glass as potential biodegradable metal. J Biomed Mater Res B Appl Biomater 100B(2):368–377
68. Zhang XL, Chen G et al (2012) Mg-based bulk metallic glass composite with high bio-corrosion resistance and excellent mechanical properties. Intermetallics 29:56–60
69. Waterman J, Staiger MP (2011) Coating systems for magnesium-based biomaterials—state of the art. Paper presented at the Magnesium Technology
70. Chen XB, Birbilis N et al (2011) A review of corrosion resistant conversion coatings for magnesium and its alloys. Corrosion 67(3):1–16
71. León B, Jansen JA (eds) (2008) Thin calcium phosphate coatings for medical implants. Springer, New York
72. Waterman J, Pietak A et al (2011) Corrosion resistance of biomimetic calcium phosphate coatings on magnesium due to varying pretreatment time. Mater Sci Eng, B 176(20):1756–1760
73. Waterman J (2012) In vitro assessment of the corrosion protection of biomimetic calcium phosphate coatings on magnesium. University of Canterbury, Christchurch, New Zealand
74. Chen XB, Yang HY et al (2012) Magnesium: engineering the surface. JOM 64(6):650–656
75. Chen X-B, Kirkland NT et al (2012) In vitro corrosion survey of Mg–Xca and Mg–3zn–Yca alloys with and without calcium phosphate conversion coatings. Corros Eng, Sci Technol 47(5):365–373
76. Chen XB, Birbilis N et al (2011) A simple route towards a hydroxyapatite–$Mg(OH)_2$ conversion coating for magnesium. Corros Sci 53(6):2263–2268

Chapter 5
Summary of Concluding Remarks

Abstract Understanding in vitro techniques (including their correct use and analysis), controlling variables appropriately during testing and proper alloying regimes are crucial to the continued development of Mg-based biomaterials. Progress in this bio-Mg area has been hampered by a lack of coordination, along with a lack of appropriate standards and insufficient cross-disciplinary collaboration. A database should be established to allow comparisons between reported tests, as well as to facilitate the creation of suitable models to predict performance of future alloys.

Keywords Database · Biomagnesium · Biomaterial · Metal · Standards · Neural network · In vitro · ASTM

5.1 Summary of Findings

5.1.1 In Vitro Experimental Techniques

The benefits and limitations of the most common biocorrosion experiments were discussed, focusing on evaluating the corrosion of Mg alloys. Each experiment has its own positive and negative attributes, and no experiment provides all the information needed to fully analyse the corrosion performance of a candidate alloy. Rather, a combination of techniques is required to obtain a more complete understanding of the corrosion taking place, both in terms of the absolute loss of material and in terms of the mechanism. The importance of understanding each technique, the relevant variables and how the choice of parameter values affects results cannot be overstated.

The corrosion test methods that have been presented in this work underpin the development of physiologically relevant in vitro techniques, which are urgently required for the prediction of in vivo biodegradation behaviour of Mg-based biomaterials. A summary of the various biocorrosion experiments investigated in this work support the critical nature of the review, and their salient features can be seen in Table 5.1.

5.1.2 Effects of In Vitro Variables on Biocorrosion of Mg Alloys

Examination of over 150 articles published on Mg biocorrosion brings to light the apparent lack of understanding of the importance of some of the most common variables for these tests [1]. For example,

1. More than 35 % of the papers performed their tests at room temperature rather than at the physiological temperature of 37 °C, affecting not only the corrosion rate but also the adsorption of proteins onto a surface [2, 3]. We have shown that depending on the alloy and test method, a change of just 17 °C results in an increase of between 50 and 800 % in the measured corrosion rate.
2. Although pH is maintained at 7.4–7.6 in the body, over 60 % of published works either did not maintain pH or completely failed to mention it at all. pH is crucial both to the binding of proteins to the surface and to corrosion, where it primarily affects the formation of layers (such as $Mg(OH)_2$ or calcium phosphate).
3. Other problems with published studies include not using a buffer to control pH (75 % of the papers), use of overly simplistic solutions (e.g. those containing only NaCl), without any of the other important inorganic minerals found in the body that affect formation of realistic corrosion layers (30 % of papers). In addition, use of a range of other simulated body fluids with varying compositions (e.g. mineral concentrations, HCO_3 levels) makes comparison between published results very challenging.

While not exhaustive, a proposed flow chart for in vitro experiment design is provided in Fig. 5.1.

5.1.3 The Need for Standards

The potential advantages of Mg over other non-resorbable biomaterials, especially for orthopaedic applications, are obvious. Fully realised, functional bioresorbable implants based upon Mg alloys would be completely unique to the field, providing the mechanical benefits of a metal combined with the degradable and biological advantages currently exhibited by polymers and other weaker synthetic biomaterials [4]. Yet, despite significant recent research, there remain many challenges to

Table 5.1 Summary of in vitro experiments, their advantages and limitations

Method	Advantages	Limitations
Unpolarised experiments		
Mass loss	• Low cost	• Provides no information on corrosion mechanisms
	• Simple to set up and perform	• Multiple samples needed for accuracy (variation between samples)
	• Easy to control environment (i.e. expose in an incubator)	• No information on time-dependant corrosion behaviour
	• Provides accurate, clearly defined data	• Multiple solution changes may be needed
	• Concurrent polarisation experiments are possible	• Corrosion product directly affects results and must be removed
Hydrogen evolution measurement	• Low cost	• Little information on corrosion mechanisms
	• Real-time measurement of corrosion	• Multiple samples needed for accuracy
	• Allows calculation of degree of alkalisation (to a varying degree of accuracy)	• Experiments with flow are difficult to measure (i.e. capture H_2)
		• Difficult to run concurrent polarisation experiments
	• Results unaffected by corrosion product	• Large number of considerations during set-up/running of test which can greatly impact results and lead to irreproducibility between tests
	• H_2 can be a problem in vivo, so its measurement is crucial	
Polarised experiments		
Potentiodynamic polarisation	• Set-up/sample prep is relatively easy	• Requires specialised equipment (potentiostat, cell, etc.)
	• Information on instantaneous corrosion rate	• Mg alloys require repolish as surface is physically altered
	• Elucidates thermodynamic differences	• Conversion to corrosion rate requires assumption of generalised corrosion. Rate measured is only "instant"
	• Illuminates differences between multiphase alloys (shifts in potential and current density)	• Reveals little about individual layers/coating for protection
	• Can be used to determine anodic/cathodic control of reactions	• Investigator variation/error can cause large differences in determined corrosion current density
	• A single sample may be used for multiple runs if reprepared	
Electrochemical impedance spectroscopy	• Set-up/sample prep is relatively easy, non-destructive	• Requires specialised equipment (ac capable potentiostat, cell, etc.)
	• Real-time measurement of corrosion resistance	• Does not reveal anodic/cathodic contributions
		• Susceptible to corrosion that occurs over time
	• Quantifies formation of any surface layers and role they play in corrosion protection over time	• Low-frequency measurement difficult for rapid corroding samples
	• Determines time-dependant breakdown of coatings	• Choice of equivalent circuit is crucial to appropriate analysis and can be difficult
	• Stern-Geary equation used to obtain approximate i_{corr} value	

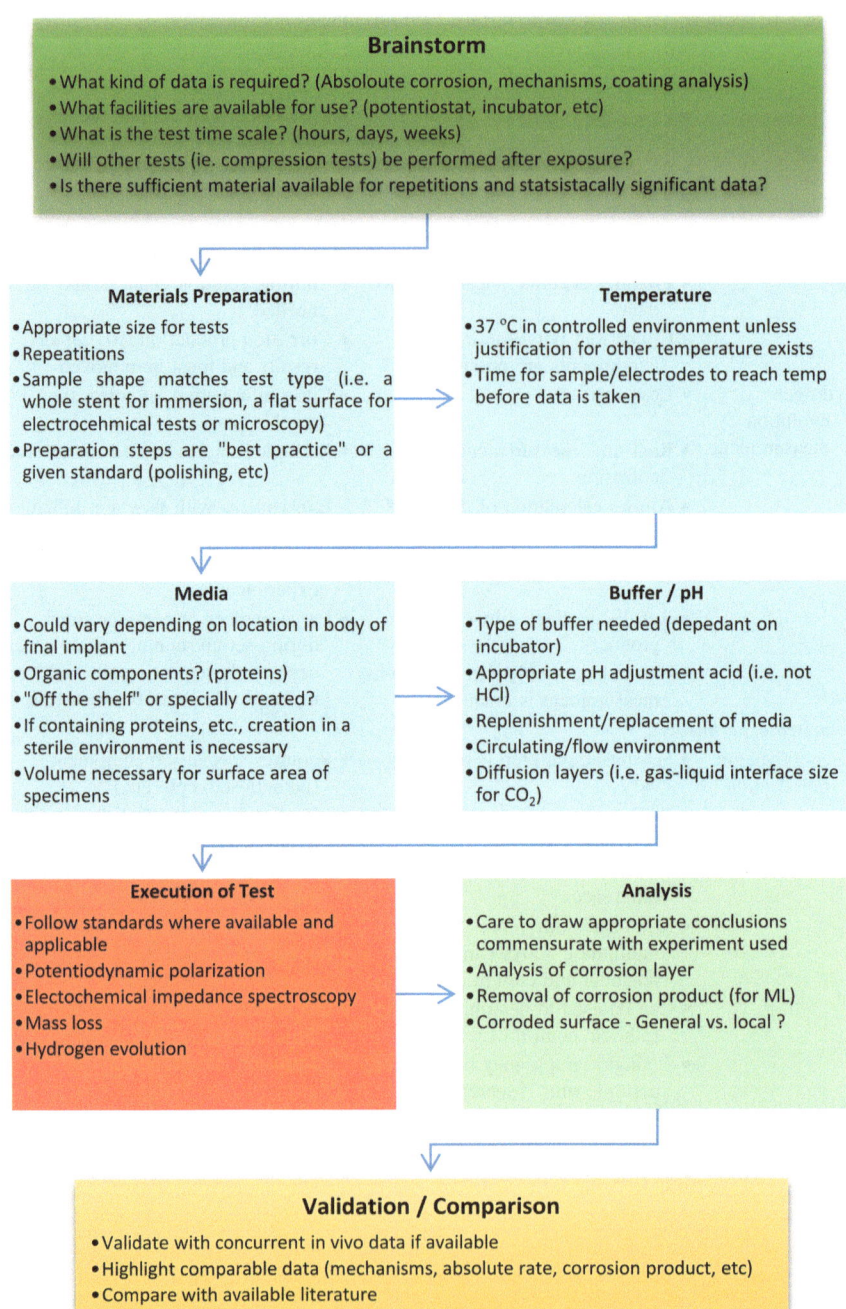

Fig. 5.1 A potential flow chart for in vitro experimental design considerations

the successful implementation of Mg-based materials in applications other than stents in humans.

As noted in previous chapters, the benefits that Mg enjoys over current implant materials, such as its biodegradability [5–10] and reduced chance of stress shielding [11], also present some of the greatest challenges to its use in a wider context. The notion that Mg alloy implants would be designed to degrade in vivo also implies that their shape and mechanical properties constantly change over the life of the implant, adding a further level of complexity to carrying out full life-cycle design and testing [12–15], while the potential harm caused by corrosion products [16] and hydrogen evolution [17, 18] is a constant concern.

The cross-disciplinary nature of biocorrosion studies unfortunately has resulted in important problems in experimental design, variable control and even analysis of results going unnoticed when studies are performed only by those who are experts in some of the these fields. Faults that cause what would otherwise be considered good science to be of limited use, or at the very least, to have to be repeated, are common in the literature, with the potential pitfalls of working with Mg only to be overcome through systematic research and careful planning.

The first step in the creation of such a systematic approach would be the establishment of guidelines for performing experiments (e.g. standards). It has already been shown that current ASTM standards for corrosion tests do not apply to Mg, especially in a biocorrosion context [19]. This has required current investigators to devise tests to the best of their ability based on their experience and any previous examples they may have come across. Although in research it is important to refrain from imposing unnecessary constraints on the types and conditions under which an experiment may be performed, the lack of standards all too often results in significant unnecessary errors, as mentioned above.

At a minimum, standards must include the following:

1. Optimal values (or ranges) for important variables such as temperature, pH, solution and buffer;
2. Experimental settings to be used, especially for electrochemical work;
3. Parameters that are required to be reported in any published work, allowing others to repeat and reproduce the work; and
4. Other core variables that should not be altered unless an experiment explicitly investigates it, such as the minimum purity level that may be referred to as "pure" Mg.

Establishment of such a set of standards should be discussed and agreed to at an international level, in a forum such as the annual Biodegradable Metals Conference [20] or World Biomaterial Congress. Input on the content and design of such guidelines needs to be taken from as many relevant international research groups, companies and government agencies as possible; for the guidelines to be of greatest use, they must be virtually universally adhered to by the biomaterial community. Any standards must not be proscriptive and in no way obstruct or discourage investigators from developing new experiments and refining those already established, but instead should provide a foundation on which most researchers could base their work.

The creation of such standards would then make it much simpler to compare research results on an international scale; it is hoped and anticipated that advances in the bio-Mg field would then develop much more rapidly and wasted time due to common experimental errors would be minimised.

A workshop was held at the FDA's Silver Spring office (30 March 2012) to discuss the latest developments in the biodegradable metals area. It was attended by 120 participants with the aim of bringing leading academics working in this field and the FDA together to discuss their views of this novel area. The outcome was an initiative to create a committee to focus on the standardisation of metals for medical use.

5.2 The Future of Mg Biomaterials

5.2.1 Development of More Realistic In Vitro Experiments in the Search for In Vivo Correlation

Due to the lack of currently available standards for Mg biocorrosion experiments, the majority of the existing literature, although occasionally promising, adds little valuable knowledge towards systematic development of the Mg alloys. The closed, static environment of typical in vitro experiments (unless special attention is paid to create a more realistic experimental set-up) will have an effect on the corrosion process itself. There is currently a disconnection between short- and long-term in vitro tests, with the latter being almost entirely unreliable. In addition, although short-term experiments are, for the most part, reproducible and accurate, their long-term relevance has not yet been established, and there is a strong need for benchmarking via parallel in vivo studies. In conjunction with this, investigation of mechanical performance features are considered critical to the success of an Mg-based implant, including stress corrosion cracking and corrosion fatigue, may only be adequately determined via long-term experiments. Although the mechanistic insights and estimates of dissolution provided by short-term tests are certainly useful, longer time-frame experiments will be necessary as part of the "next step" in the development of Mg biomaterials.

One factor that may have a large influence on such longer-term experiments is the flow of media around the implant, with different rates mimicking various situations in the body. The movement of biological fluids around an implant is an area of study that has received relatively little attention in the design of in vitro tests for Mg alloys. It has been shown in the literature that flow can have a considerable effect on biocorrosion of Mg [21, 22]. However, faster flow rates do not automatically result in increased corrosion [23]. Flow of the body's fluids can affect bone adsorption and consequently likely plays a vital role in the success or failure of a given implant [24]. Published Mg biodegradation studies have used a number of methods to induce flow, including shaking platforms [21], rotating electrodes [23, 25] and sheer-stress-controlled flow parallel to the sample surface [22, 26]. Levesque et al. created what is perhaps the most realistic flow cell for degradable metallic biomaterials to date [22]. However, their set-up was designed only for stent applications in coronary arteries rather than orthopaedic applications for which significantly slower flow rates are observed.

Such flow rates may be achieved via the use of bioreactors, which can offer near-ideal test conditions, mimicking those encountered in many orthopaedic applications. They are a controlled environment in which variables crucial to Mg corrosion, such as temperature and pH, may be monitored and carefully controlled. There are currently available, well-established procedures for the use of such systems, and these procedures could be adapted for Mg materials. A bioreactor specifically for Mg (or other metallic biomaterials) could also be designed to incorporate electrochemical measurements that could be carried out while the experiment is running. A range of flow conditions that may be encountered in the body could be simulated, while the chemical reactions occurring at the surface could be analysed. Additionally, this would allow easier (and arguably more realistic) study of protein and cell interactions with the samples, as the adsorption and desorption of these organic components is affected by flow and induced sheer stress at the surface [27]. Such conditions would be more applicable and comparable to what occurs in the body than the current static cell culture experiments that have been performed on Mg biomaterials.

5.2.2 Effect of Organic Compounds on Mg Biocorrosion

The effect of amino acids and proteins on the corrosion of Mg remains unclear. Consequently, at this stage, it is not possible to recommend how these organic compounds should be used in the in vitro prediction of Mg implant performance, other than their use in established toxicity experiments. In this book, attempts were made to determine some of the more basic characteristics of the interactions between proteins and Mg alloys using electrochemical techniques. However, alternate methods, such as quartz crystal microbalance (QCM), show significant promise although there are difficulties to overcome. For QCM, issues include the coating method for the crystals, allowing slower Mg corrosion rates (as the authors found sputter-coated and purchased Mg QCM crystals to degrade faster than measurements could record). Attenuated total reflectance Fourier transform infrared spectroscopy (ATR-FTIR) may help detect protein attachment without modification of the surface after immersion. Atomic force microscopy (AFM) may also detect the attraction of proteins to the Mg surface, although for all the above methods, the corrosion of Mg remains an issue. An outline of current tests available to investigate protein and cellular attachment is provided in Appendix B.

A more comprehensive study of the effect of organic compounds on Mg corrosion will require a multidisciplinary approach based on biomolecular chemistry and electrochemistry. Although not yet available in the literature, it is hoped that such work will be carried out and published in the near future.

5.2.3 Establishment of a Corrosion Database and Neural Networks

Even though, by the time of this publication, there will likely be well over 200 published works investigating the biocorrosion of a wide range of Mg alloys, there has been little to no effort to collect this data for comparison and wider dissemination. The list presented in this book in Appendix C goes some way to compiling such a record, but more work is needed to make this information of wider use. The establishment of a database of collected results would allow more detailed comparisons and allow for greater and more accurate discussion of new data in the light of what has already been published.

After a critical amount of data has been obtained, this database could eventually lead to the development of a tool, such as a neural network (discussed previously [12]), which would allow insight into the performance of a potential material before it has been physically produced and tested. A neural network essentially creates advanced algorithms based on inputs (such as alloying levels) and their respective outputs (including corrosion rate and mechanical properties). Once created, this allows the investigator to alter whatever input variable they choose and predict the resulting performance. The beauty of such a technique is that it becomes increasingly accurate as more physical data are entered. A tool such as this would be extremely valuable for the final patient-specific implant design, where each implant (material and coating) could theoretically be designed explicitly to match the needs of the intended recipient.

The authors have already started to construct such a database, but it is hoped that through collaboration with the wider community, the development of the database will accelerate, and it will become more useful.

5.3 Conclusions

As we reach the end of what can perhaps be considered the first decade of modern research into Mg as a biomaterial, critical appraisal of the open literature suggests that it is now time for researchers in this field to transition from simply appreciating the potential benefits of Mg towards actually engineering designs into a final product. To do this, a far more systematic, interdisciplinary approach is needed. A greater understanding of Mg bioperformance parameters will lead to the establishment of appropriate standards, which will in turn set in motion accelerated research and collaboration on an international level.

Other key activities which will lead to even greater research advances in the development of Mg's potential as a viable implant material include the following:

- Establishing a database of published and unpublished results with their associated experimental parameters. Such a record would help lead to the development of an alloy performance prediction method (such as a neural network). Such a database should be open access for greatest impact.
- Determining the toxicity of all potentially hazardous alloying elements. This needs to be further explored before these materials could be seriously considered as biomaterial candidates; in particular, elements such as the family of rare earths (Ce, La, Pr, Nd, Y, Gd), Li and Zr must receive greater attention, along with their long-term exposure hazards.
- Developing coatings for Mg biomaterials. Appropriate coatings will likely be necessary for any successful Mg implant to provide initial protection by slowing hydrogen evolution and permitting more ready implant acceptance in vivo. Any such coating, however, must display a controlled biodegradation rate.

It is unlikely (and both untenable and unethical) that an extremely large number of in vivo experiments can be carried out to determine the biomaterial's performance of the wide range of Mg alloys whose basic properties (such as long-term toxicity and corrosion behaviour) have not yet been adequately explored in vitro. As more in vitro knowledge is obtained, there will be a transition towards more in vivo tests. Regardless, extreme caution must always be taken to ensure that the value of in vivo tests is maximised by understanding the alloy performance using appropriate in vitro methods.

References

1. Kirkland NT, Birbilis N et al (2012) Assessing the corrosion of biodegradable magnesium implants: a critical review of current methodologies and their limitations. Acta Biomater 8(3):925–936
2. Omanovic S, Roscoe SG (1999) Electrochemical studies of the adsorption behavior of bovine serum albumin on stainless steel. Langmuir 15(23):8315–8321

3. Jackson DR, Omanovic S et al (2000) Electrochemical studies of the adsorption behavior of serum proteins on titanium. Langmuir 16(12):5449–5457
4. Staiger MP, Pietak AM et al (2006) Magnesium and its alloys as orthopedic biomaterials: a review. Biomaterials 27(9):1728–1734
5. Saris N-EL, Mervaala E et al (2000) Magnesium: an update on physiological, clinical and analytical aspects. Clin Chim Acta 294(1–2):1–26
6. Vormann J (2003) Magnesium: nutrition and metabolism. Mol Aspects Med 24:27–37
7. Merck International (2006) Water, electrolyte mineral, and acid/base metabolism. In: Porter RS, Kaplan JL (eds) Merck manual of diagnosis and therapy. Merck & Co., Inc
8. Okuma T (2001) Magnesium and bone strength. Nutrition 17:679–680
9. Wolf FI, Cittadini A (2003) Chemistry and biochemistry of magnesium. Mol Aspects Med 24:3–9
10. Hartwig A (2001) Role of magnesium in genomic stability. Mutat Res Fundam Mol Mech Mutagen 475:113–121
11. Department TAFST (2006) Magnesium alloys. The American Foundry Society, Schaumburg, Illiniois
12. Kirkland NT, Staiger MP et al (2011) Performance-driven design of biocompatible Mg-alloys. JOM 63(6):28–34
13. Schinhammer M, Hänzi AC et al (2010) Design strategy for biodegradable Fe-based alloys for medical applications. Acta Biomater 6(5):1705–1713
14. Adachi T, Osako Y et al (2006) Framework for optimal design of porous scaffold microstructure by computational simulation of bone regeneration. Biomaterials 27(21):3964–3972
15. Hollister SJ, Maddox RD et al (2002) Optimal design and fabrication of scaffolds to mimic tissue properties and satisfy biological constraints. Biomaterials 23(20):4095–4103
16. Witte F, Ulrich H et al (2007) Biodegradable magnesium scaffolds: part 1: appropriate inflammatory response. J Biomed Mater Res Part A 81(1):748–756
17. Song G (2007) Control of biodegradation of biocompatible magnesium alloys. Corros Sci 49(4):1696–1701
18. Seal CK, Vince K et al (2009) Biodegradable surgical implants based on magnesium alloys: a review of current research. IOP conference series: materials science and engineering, vol 4, p 012011
19. Witte F, Nellesen J et al (2006) In vitro and in vivo corrosion measurements of magnesium alloys. Biomaterials 27(7):1013–1018
20. Biodegradable Metals Conference (2012). http://www.biometal2011.org/. Accessed 31 Mar 2012
21. Yang JX, Cui FZ et al (2009) Characterization and degradation study of calcium phosphate coating on magnesium alloy bone implant in vitro. IEEE Trans Plasma Sci 37(7):1161–1168
22. Levesque J, Hermawan H et al (2008) Design of a pseudo-physiological test bench specific to the development of biodegradable metallic biomaterials. Acta Biomater 4(2):284–295
23. Hiromoto S, Yamamoto A et al (2008) Polarization behavior of pure magnesium under a controlled flow in a NaCl solution. Mater Trans 49(6):1456–1461
24. Johansson L, Edlund U et al (2009) Bone resorption induced by fluid flow. J Biomech Eng 131(9):094505
25. Hiromoto S, Yamamoto A et al (2008) Influence of pH and flow on the polarisation behaviour of pure magnesium in borate buffer solutions. Corros Sci 50(12):3561–3568
26. Chen Y, Zhang S et al (2010) Dynamic degradation behavior of MgZn alloy in circulating M-SBF. Materials Letters 64
27. McIntire LV, Wagner JE et al (1998) Effect of flow on gene regulation in smooth muscle cells and macromolecular transport across endothelial cell monolayers. Biol Bull 194(3):394–399

Appendix A
Preparation of Kirkland's Biocorrosion Media

Cleaning of Glassware

- Use only new glassware
- All bottles, flasks, beakers and other glassware should be cleaned with distilled water 3–4 times. Following this, they should be washed with high-purity ethanol (99.8 %) 2–3 times.
- If required, immerse bottles in dilute HCl solution for 4–6 h. Remove from solution and repeat above steps.

Creation of Media Buffer

- All dissolution of chemicals should take place in a flow hood to minimise dust pickup.

1. Pour 900 ml of distilled water into a 1-L beaker. Place a clean magnetic stirrer bar in the beaker and place on a heated magnetic plate.
2. Adjust heated plate so the temperature of bulk solution is 37 °C ±1.
3. Add the chemicals given in Table A1 one by one in the order provided. Ensure each is completely dissolved before adding the next.
4. Buffer choice depends on the intended experimental atmosphere. Both may be added if desired. Add the buffer (Table A2) in 1 g amounts to minimise localised changes in pH.
5. Finally, add distilled water to the beaker until 1 L is reached ($\sim$95 ml).

pH Adjustment for HEPES Buffer

6. Before any pH measurement, calibrate the pH meter to a minimum of two points (*e.g.* 7 and 10).
7. Recheck the temperature of the solution to ensure it is 37 °C ±1.
8. Measure the pH of the KBM, and add 1 M NaOH in 0.25 mL portions until a pH of 7.4 is reached.

N. T. Kirkland and N. Birbilis, *Magnesium Biomaterials*, SpringerBriefs in Materials, DOI: 10.1007/978-3-319-02123-2, © The Author(s) 2014

Table A1 Components and concentrations of KBM

Compound	Amount (g/L)
NaCl	5.4
D-Glucose ($C_6H_{12}O_6$)	0.9
KCl	0.38
$CaCl_2$	0.28
Na_2HPO_4 (Anhydrous)	0.122
$MgSO_4$	0.06
Phenol Red ($C_{19}H_{14}O_5S$)	0.011

Table A2 Buffers and concentrations for addition to KBM

Buffer	Amount (g/L)
SB ($NaHCO_3$)	2.2
HEPES ($C_8H_{18}N_2O_4S$)	5.96

Table A3 Buffers and concentrations for addition to KBM

Component	Human plasma	KBM
Na	142	120.3
Cl^-	103	102.5
K^+	5.0	5.0
Ca^{2+}	2.5	2.5
Mg^{2+}	1.5	1.5
HPO_4^{2-}	1.0	0.9
SO_4^{2-}	0.5	0.5
D-Glucose	5	5
Bicarbonate (HCO_3)	22–30	±26.2
HEPES	–	±25
Phenol Red	–	0.031

pH Adjustment for NaHCO$_3$ Buffer

6. The concentration of CO_2 in the incubator will determine the pH of the media.
7. Adjust the volume % CO_2 of the incubator to 5 %, and allow solution to stabilise over 4 h. Measure the pH of the media and raise CO_2 level if pH is too high (and lower CO_2 level if pH is too low).
8. Let it settle again for 4 h, and measure pH again. Typically, 4.7–5.1 % CO_2 results in a pH of 7.4.

Storage

– Store in a refrigerator at 5–10 °C, use within 2 weeks (Table A3).

Appendix B
Experimental Techniques to Detect Protein and Cellular Adhesion

Test	Use	Procedure	Comments
X-ray photoelectron spectroscopy (XPS)	Elemental composition	Surface is irradiated by beam of monochromatic x-rays, released photoelectrons are captured	Common, non-destructive, not useful for multiprotein studies
Secondary ion mass spectroscopy (SIMS)	Elemental and molecular composition	Bombardment of surface with focused beam of ions/atoms resulting in emission of secondary particles	Common, good protein resolution
Time-of-flight SIMS (ToF-SIMS)	Elemental and molecular composition	Similar to SIMS, secondary ions are accelerated through field-free drift region, which separates heavy/light ions	Greater resolution, increased sensitivity
Fourier transform infrared spectroscopy (FTIR)	Chemical bonds and conformation	IR beam is reflected from a surface, and spectrum is developed from IR absorbance at different frequencies	Common, limited for multiprotein studies (complicated spectra), quick
Attenuated total frequency (ATR-FTIR)	Chemical bonds and conformation	Similar to FTIR, total internal reflection of IR beam back into detector along the region of contact with sample surface	Reduced surface area required than FTIS
Solute-depletion test	Amount adsorbed	Changes in bulk concentration of protein in a solution are measured over time	Assumes all protein loss is due to material attachment
Protein labelling, staining and florescence	Amount adsorbed and areas	Proteins labelled with fluorescent or radioactive probes, concentration of adsorbed protein deterred on surface	Easy to perform
Immunoassay (eg. ELISA)	Presence and bioactivity	An antibody is directed against a protein, bonding to specific portions of the target protein	Common

(continued)

N. T. Kirkland and N. Birbilis, *Magnesium Biomaterials*, SpringerBriefs in Materials, DOI: 10.1007/978-3-319-02123-2, © The Author(s) 2014

(continued)

Test	Use	Procedure	Comments
Ellipsometry	Layer thickness	Polarised light directed at surface at an angle, reflected back to detector. Changes in reflected light provide information on interface	Common, nanometer resolution
Surface plasmon resonance (SPR) and Neutron reflectivity (NR)	Layer thickness	Similar to ellipsometry, but uses excitation of surface plasmons and beam of neutrons, respectively	NR is not widely used due to equipment and potential radiation problems
Quartz crystal microbalance (QCM)	Protein adsorption over time	Measure change in frequency for coated quartz crystal as a solution comes into contact	Real-time measurement of protein attachment, quantitative values
Matrix-assisted laser/desorption/ ionisation time-of-flight (MALDI-ToF)	Protein identification	Matrix molecules are added to surface prior to irradiation with pulse laser	Difficult sample preparation, matrix fluid required, low detection limits
Scanning and transmission electron microscopy (SEM/TEM)	Elemental and layer analysis	Surface analysed under electron microscope	Sample preparation (drying, coating) may cause serious issues, poor quantitative analysis
Atomic force microscopy (AFM)	Protein attraction, layer property	"Feels" surface with a very sharp microscale cantilevered tip	Allows tests in situ, can be very hard to set up
Scanning tunnelling microscopy (STM)	Protein attraction, layer property	Measures quantum mechanical tunnelling of current between conductive surface and very sharp metallic tip.	Can capture extremely fine changes in height, lateral information not available from AFM
Surface-enhanced infrared adsorption spectroscopy (SEIRA)	Protein functionality	IR technique, amplifies received signal by two orders	Not widely used

Appendix C
Categorized Collection of In Vitro and In Vivo Studies on Magnesium Biomaterials

The following table is provided as a powerful resource and supplementary information for the Springer Briefs publication, "Magnesium Biomaterials—Design, Testing and Best Practice" by Nicholas T. Kirkland and Nick Birbilis. Great effort has been made to ensure its completeness and accuracy, but errors may remain.

Term	Abbreviation
Solution	
Artificial plasma	AP
Cell culture	CC
Deionized water	DiO
Earle's balanced salt	EBSS
Hank's balanced salt solution	HBSS
McCoys culture solution	McC
Minimum essential medium	MEM
CModel saliva	MS
Phosphate buffer solution	PBS
Sodium Chloride-based solution	NaCl
Substitute sea water	Sea
Tyrode's solution	TS
Unnamed simulated body fluid	uSBF
Buffer	
Borate buffer	BB
HEPES	HPS
Sodium Bicarbonate and $x\%$ CO_2 environment	SB_{CO2}
Coating/prework	
Anodized	And
Bulk metallic glass	BMG
Calcium phosphate coating	CaP
Cast	Cst
Dynamic environment	Dyn
Extruded	Ext
Heat treated	HT

(continued)

N. T. Kirkland and N. Birbilis, *Magnesium Biomaterials*, SpringerBriefs in Materials, DOI: 10.1007/978-3-319-02123-2, © The Author(s) 2014

(continued)

Term	Abbreviation
Microarc oxidation	μArc
Rolling	Rlg
Rotating disk	RtD
Solution treated	ST
Experiments	
3-Point bend test	3PB
Alkaline Phosphatase test	APT
Atomic absorption	AA
Chemotaxis assay	CA
Compression test	CT
Confocal microscopy	CfM
Cytotoxicity	CTx
Dendritic cell viability	DCV
Energy-dispersive X-ray spectroscopy	EDS
Fatigue test	Ftg
Haemolyses	Hml
Hydrogen evolution test	H2E
In vivo experiments	In Vivo
Kinetic clotting time test	KCT
Laser profilometer	LP
Linear polarisation	LP
Mass loss	ML
Mg Ion measurement	Ion
Microcomputed topography	μCT
Microcell corrosion test	μCell
Mixed leukocyte reaction	MLR
Nano indentation	NI
Optical microscopy	Opt
pH recording	pHR
Platelet adhesion	PA
Potentiodynamic polarisation	PDP
Scanning electron microscopy	SEM
Slow strain rate test	SSRT
Tensile test	TT
Total protein assay	TPA
Vickers hardness test	VH
Volume fraction measurement	VfM
Wettability	Wet
X-ray diffraction	XRD

Important Points to Note:

- Amounts of alloying elements assume the remainder is Mg
- For pH Control, "*Y*" indicates control and "*N*" indicates no control
- "–" indicates the variable was not stated or performed.

Alloy	Solution	Buffer	Coating/prework	°C	pH control	Time	Experiments	References
"Pure" Mg, 0.6Ca, 0.8Ca, 1Ca, 1.2Ca	CC + 10 % FBS	–	–	37	Y, 7.4	3, 6 days	DCV, Ion, MLR, CA	[1]
99.5 % Mg	MS	–	±PCL	–	Y, 7.36	2, 3 weeks	SEM, ML, Opt, (In Vivo)	[2]
99.8 % Mg, 0.5Ca, 0.8Ca, AZ21	MEM + 15 % FBS	SB_{CO2}	–	37	–	8 days	ML, CTx, TPA, APT, SEM, EDS, CfM	[3]
99.87 % Mg, 5Ca, AZ91, AM50	MS	–	Sand/Mould Cst, Ext	20	Y, –	164 h	ML, VH, TT, Opt, SEM, EDS	[4]
99.9 % Mg	HBSS	–	Na_2CO_3/NaHCO$_3$	25	N (12)	75 days	ML, SEM, EDS, XRD	[5]
99.9 % Mg	NaCl (±HPS, NaHCO$_3$), EBSS, MEM (±10 % FBS)	HPS	–	37	Y, 7.4	14 days	Ion, EDS, Opt	[6]
99.9 % Mg	HBSS	–	HT	20	–	625 h	SEM, XRD, EDS,	[7]
99.9 % Mg	0.6 % NaCl	–	RtD	37	N, 6	6 h	PDP, OCP, EIS, Opt,	[8]
99.9 % Mg	0.6 % NaCl	BB	RtD	37	Y, 6.7, 7.6, 9.3	6 h	EIS, OCP, XPS, Opt	[9]
99.9 % Mg	PSS	–	Foam	37	–	144 h	ML, pHR, CT, Opt, SEM, XRD	[10]
99.9 % Mg	uSBF	HPS	ST	37	Y, 7.4, 9	5 days	CTx, Wet, LP, SEM, EDS	[11]
99.9 % Mg	uSBF	–	±CaP	37	N, 7.4	21 days	ML, pHR, AA, SEM, EDS, XRD	[12]

(continued)

(continued)

Alloy	Solution	Buffer	Coating/prework	°C	pH control	Time	Experiments	References
99.9 % Mg	uSBF	–	Annealed	37	N, 7.6	21 days	ML, pHR, Opt, SEM, EDS, XRD	[13]
99.9 % Mg	uSBF (±Cl)	–	ST/HT	37	N	14 days	ML, pHR, CTx, SEM, XRD, XPS	[14]
99.9 % Mg, AZ31, AZ61, AZ91	HBSS (±Ca, Mg)	–	½ Oxidised	20	–	1080 h	ML, SEM, XRD, Opt	[15]
99.92 % Mg	0.9 % NaCl	–	Ti Coating	25	–	–	PDP, SEM, EDS, XRD	[16]
99.946 % Mg, 99.967 % Mg, 99.993 % Mg	HBSS	–		20	N, 7.4	180 h	H2E, ICP, SEM, EDS, XPS	[17]
99.947 % Mg, AZ91	uSBF	HPS	–	37	7.4	–	PDP, EIS, SEM	[18]
99.95 % Mg	PBS, DiO, McC + 5 % FBS	–	–	37	–	5 days	ML, OCP, CTx, EIS, EDS, SEM	[19]
99.95 % Mg, 99.9999 % Mg, AZ31, AZ91	1 % NaCl	BB	–	25	Y, 6.5, 9	168 h	ML, pHR	[20]
99.95 % Mg, ZX152	HBSS (±Ca, Mg)	–	–	37	–	4 days	H2E, SEM, ICP	[21]
99.96 % Mg	uSBF	HPS	ST/HT	37	N, 7.4	40 days	ML, pHR, SEM, EDS, XRD, ICP	[22]
99.96 % Mg	HBSS	–	pH Test	37	Y, 5.5–8	7 days	H2E, Ion, PDP, EIS, Opt, SEM	[23]
99.96 % Mg	HBSS	–	Conversion in HF	37	–	2 days	EIS, PDP, SEM, AFM, XRD, XPS	[24]

(continued)

(continued)

Alloy	Solution	Buffer	Coating/prework	°C	pH control	Time	Experiments	References
99.96 % Mg	HBSS	Citric	Steric Acid Coating	37	Y, 7.4	24 h, 80 d	PDP. EIS, Opt, SEM, FTIR, XRD	[25]
99.96 % Mg	HBSS	TRIS	±Sol–Gel Coating	37	Y, 7.4	15 days	EIS, SEM, EDS, XRD	[26]
99.96 % Mg, 0.6Ca, 1.2Ca, 1.6Ca, 2Ca	uSBF	–	HT	20	–	–	PDP, SEM, Opt, XRD, 3PB	[27]
99.96 % Mg, AZ91	HBSS	–	±And	37	N, 6	30 days	PDP, EIS, H2E, pHR, Opt	[28]
99.976 % Mg, AZ31, AZ61, AZ91	m-SBF	HPS	–	37	N, 7.4	24 days	ML, PDP, EIS, SEM, EDS	[29]
99.98 % Mg, 1Ca, 1Zn, 12Li, AZ31,	0.1M NaCl	–	–	–	N, 7	22 days	ML, OCP, LP, Opt, XRD	[30]
99.98 % Mg, LAE442, AZ31	5,10,35 g/L NaCl, PBS ±0.1/1/ 10 g/L BSA	–	–	–	–	5 m	PDP, EIS, µCell, Opt, SEM, EDS	[31]
99.98 % Mg, LAE442, AZ31	0.9 % NaCl, PBS ±1/10 g/L BSA	–	–	–	–	–	PDP, EIS, Opt, SEM, EDS, µCell	[32]
99.99 % Mg	HBSS	–	–	37	Y, 7.5	30 days	ML, H2E, KCT, pHR, SEM, EDS (+ In Vivo)	[33]
99.99 % Mg, 0.8Ca, 1Zn, 1Mn, 1.34Ca-3Zn, AZ31	EBSS, MEM ± 40g/L BSA	SB_{CO2}	–	37	Y	7,14,21 days	ML, phR, SEM, (+In Vivo)	[34]

(continued)

(continued)

Alloy	Solution	Buffer	Coating/prework	°C	pH control	Time	Experiments	References
99.99 % Mg, 6Zn	uSBF	–	ST	37	N, 7.44	3, 30 days	PDP, EIS, ML, TT, CT, ICP-AES, SEM, EDS, XRD (+In Vivo)	[35]
99.998 % Mg, 1Nd, 0.05Sn, 1Ce, 0.4Ca, 1.34Ca, 16.2Ca, 3 Zn, 6.2Zn, 10Zn, 3Al, 4Ce, 0.5Y, 0.1Sr, 1La, 2Y, 4La, 4Nd, 5Ca, 0.4Ca-3Zn, 0.4Ca-10Zn, 7Al-0.05 Ti AZ31, ZE41, AE44, AZ91	MEM	SB_{CO2}	–	37	N, 7.4	14 days	ML, PDP, SEM,	[36]
AM50	HBSS	–	PLA	37	–, 7.4	15 days	PDP, CTx, Opt, SEM, EDS	[37]
AM50	HBSS	–	PCL	37	–	20 days	PDP, TEM, EDS, Opt, 3PB, CTx, SEM	[38]
AM50	HBSS	–	PVA	37	–, 7.4	1–60 days	PDP, TEM, EDS, XRD, CTx, SEM, FTIR, Wet	[39]
AX53	0.9 % NaCl (±0.7 g/l NaHCO3), HBSS	–	–	20	–	6, 24 h	PSP, ML, SEM, EDS, XRD	[40]
AZ31	uSBF	–	CaP, Shaken	37	Y, 7.3	3, 5, 7, 11, 15 days	ML, SEM, EDS, XRD	[41]
AZ31	3 % NaCl	–	CaP	37	–	15 days	ML, TT, SEM, EDS, XRD, FTIR	[42]

(continued)

(continued)

Alloy	Solution	Buffer	Coating/prework	°C	pH control	Time	Experiments	References
AZ31	8 g/l NaCl, PBS	–	HT/Rolled / ECAP	20	–	6 days	OCP, PDP, EIS, SEM, EDS, FTIR	[43]
AZ31	HBSS	–	–	37	Y, 7.5	30 days	ML, H2E, SEM, EDS (+In Vivo)	[44]
AZ31	uSBF	–	–	37	Y, –	72 h	EIS, Opt, SEM, EDS, XPS	[45]
AZ31	HBSS	–	Squeeze Cst, Hot Rolled, ECAP	–	Y, 7	20 days	ML, EIS, Ftg, SEM, TEM, Opt	[46]
AZ31	HBSS	–	Squeeze Cst, Hot Rolled, ECAP	–	Y, 7	20 days	ML, Opt, TEM	[47]
AZ31	uSBF	–	MAO	37	Y, 7.25	28 days	EIS, PDP, SEM, XRF	[48]
AZ31, AZ61, AZ91	HBSS (±Ca, Mg)	–	½ Oxidised	20	–	1080 h	ML, SEM, XRD	[49]
AZ31, AZ91, WE43	0.9 % NaCl	–	µArc	20/37	–	90 m	OCP, PDP, Opt	[50]
AZ31, LAE442	0.5–3.5 % NaCl, PBS (±0–10 g/l BSA), McC	HPS	–	20	–	–	PDP, EIS, µCell	[51]
AZ31B	uSBF	SB$_{CO2}$	Fluoride treatment	37	Y, 7.4	168 h	PDP, pHR, 3PB, SEM, EDS, XRD	[52]
AZ61 + 0.4Ca, AZ91 ± 1Ca	uSBF	HPS	–	37	7.4	–	PDP, EIS, SSRT, SEM, EDS	[53]

(continued)

(continued)

Alloy	Solution	Buffer	Coating/prework	°C	pH control	Time	Experiments	References
AZ63	0.9 % NaCl, TS	–	ST/HT	37	Y, 7.2	14 days	ML, OCP, pHR, Opt, AA	[54]
AZ91	uSBF (±1 g/l BSA)	HPS	–	37	Y, 5, 7.2	7 days	ML, OCP, PDP, EIS, SEM, EDS, FTIR	[55]
AZ91	uSBF	HPS	Sand Cst	37	7.4	–	PDP, SSRT, ICPAES, SEM	[56]
AZ91	uSBF	–	–	37	–	35 h	EIS, PDP, SEM	[57]
AZ91	0.9 % NaCl, HBSS, ±FBS	–	–	37	Y	–	LP, PDP, EIS	[58]
AZ91	uSBF	–	Polymer Coating	37	N, 7.4	2 months	PDP, ML, CTx, pHR, ICPMS, SEM	[59]
AZ91	uSBF	–	Zr Coating	–	–	–	PDP, EIS, SEM, XRD, XPS	[60]
AZ91	uSBF	–	Al₂O₃ Coating	–	–	18 h	PDP, EIS, XPS, SEM	[61]
AZ91	uSBF	–	–	–	–	1, 4, 7 days	ML, H2E, OCP, EIS, Opt, SEM, EDS, FTIR, XPS	[62]
AZ91 + 20% HA	Sea, MEM (±10 % FBS)	–	–	37	–	24, 72 h	PDP, EIS, ML, VH, NI, CTx, Opt, SEM, EDS, XRD	[63]
AZ91, WE43, AM60	HBSS	HPS	Dyn	37	Y, 7.3–7.5	168 h	AA, SEM, EDS, XRD, EPMA, FTIR, Opt	[64]

(continued)

(continued)

Alloy	Solution	Buffer	Coating/prework	°C	pH control	Time	Experiments	References
AZ91D	uSBF	–	CaP	37	N, 7.4	48 h	PDP, EIS, SEM, EDS, XRD	[65]
AZ91D	uSBF	–	HT	37	7.3–7.4	168 h	ML, PDP, SEM, EDS, Opt	[66]
AZ91D	uSBF		MAO	37	–, –	28 days	PDP, SEM, XRD, Opt	[67]
AZ91D	uSBF	HEPES	–	37	–, –	–	SSRT, TT, SEM	[68]
AZ91D, LAE442	Sea	–	–	20	–	240 h	PDP, SEM, EDS (+In Vivo)	[69]
CaMgZn BMG, HP Mg	MEM	SB_{CO2}	–	37	Y, 7.4	24 days	ML, H2E, PDP, XRD, DSC, VH, Opt, SEM, EDS	[70]
CP Mg, HP-Mg, 1Zn, 2Zn-0.2Mn, ZE41, AZ91	HBSS	–	±And	37	–	30 days	H2E	[71]
HP Mg, AZ31	HBSS	–	And, Dyn	–	Y, 7	20 days	ML, EIS, Opt	[72]
Mg BMG	uSBF,	–	–	37	–	30 days	pHR, PDP, CTx, SEM, XPS CT, Cell Test, Opt	[73]
Mg BMG	PBS	–	–	37	7.4	–	PDP, OCP, SEM, EDS, XRD, Opt	[74]
Mg BMG	uSBF	–	–	–	N, 7.3–7.4	4 days	H2E, PDP, EIS, OCP, SEM, EDS, XRD	[75]

(continued)

(continued)

Alloy	Solution	Buffer	Coating/prework	°C	pH control	Time	Experiments	References
Mg-(1–5)Zr-(0–5)Sr	uSBF, MEM	–	–	37	–,–	–	PDP, H2E, CT, CTx, MTS, Hml, Opt, XRF, SEM, EDS, XRD (In Vivo)	[76]
Mg-0.4Ca, 0.6Ca, 0.8Ca, 1Ca, 1.5Ca, 1.7Ca, 2Ca, 2.5Ca, 3Ca, 4Ca	0.05 – 5 % NaCl	TRIS	Ext	21	Y, 7.4	900 h	OCP, PDP, H2E, TT	[77]
Mg-0.4Ca, 0.6Ca, 0.8Ca, 1Ca, 1.5Ca, 1.7Ca, 2Ca, 2.5Ca, 3Ca, 4Ca	5 % NaCl	TRIS	Ext	21	Y, 7.4	900 h	OCP, PDP, H2E, TT	[78]
Mg-0.4Ca, 0.8Ca, 2Ca, AZ91	5 % NaCl	–	±HT	–	–	72 h	ML	[79]
Mg-0.6Ca, 1.2Ca, 1.6Ca	uSBF	–	HT, Ion Implanted (Zinc)	20	–	–	PDP, SEM, EDS, NI	[80]
Mg-0.8Ca	–	–	–	–	–	2, 4, 6, 8 weeks	ML, SEM, EDS, Pull Out Test, μCT, (In Vivo)	[81]
Mg-0.8Ca, 5Ca	HBSS	–	–	–	–	4 h	H2E, XRD,	[82]
Mg-1.2Mn-1Zn	PRMI1640	SB$_{CO2}$	CaP	37	–	5 days	CTx, SEM, XRD, XPS, EDS	[83]
Mg-1.2Mn-1Zn	PRMI1640	SB$_{CO2}$	CaP	37	Y, 7.4	5 days	CTx, Cell Count, SEM, XRD, XPS	[84]
Mg-1.2Mn-1Zn	0.9 % NaCl	–	CaP	37	N, 7.4	1, 2, 4, 9 d	PDP, SEM, EDS, SAXS, XRD, XPS	[85]

(continued)

(continued)

Alloy	Solution	Buffer	Coating/prework	°C	pH control	Time	Experiments	References
Mg1.5RE, MgRE₁	PBS	–	–	–	–	1, 2, 3 days	PDP, H2E, SEM, TEM, XRD, EDX	[86]
Mg-10Dy, Mg	MEM$_{FBS}$	SB$_{CO2}$	–	37	N, 7.4	3–28 days	ML, Opt, SEM, EDS, XPS, XRD, CTx	[87]
Mg-1Ca, 1Zn, 12Li; AZ31	0.1 M NaCl	-	–	20	N	22 days	ML, OCP, PDP, XRD	[88]
Mg-1Ca, 2Ca, 3Ca	uSBF	–	Rlg / Ext	37	N, 7.4	250 h	H2E, pHR, PDP, CTx, EDS, XRD, Opt, TT (+ In Vivo)	[89]
Mg-1Ca, 5Ca, 10Ca	MEM	–	Powder Metallurgy	37	–	1,3,12, 72 h	EIS, PDP, OCP, TT, Ion, CTx, SEM, EDS, IXP-AES, XRD	[90]
Mg-1Ca, AZ31	HBSS	–	Ext, CaP	20	N, 7.4	70, 250 h	OCP, PDP, H2E, SEM, XRD, EDS	[91]
Mg-1Ca, AZ31, AZ91	HBSS, MEM ± FBS	–	–	37	–	7 days	LP, OCP, EIS, SEM, EDS,	[92]
Mg-1Mn-1Zn	HBSS, AP	–	–	–	N	288 h	ML, OCP, pHR, PDP, SEM, EDS, XRD	[93]
Mg-1Zn, 1Mn, 1Zr, 1Si, 1Sn, 1Y	HBSS, uSBF	–	–	37	–	20 days	H2E, Ion, PDP, TT, CTx, Hml, PA	[94]

(continued)

(continued)

Alloy	Solution	Buffer	Coating/prework	°C	pH control	Time	Experiments	References
Mg-1Zn-0.2Mn	uSBF	–	±PTMC	37	N, 7.4	30 days	Flow ML, PDP, EIS, SEM, Hml, CTx, XRD, Opt	[95]
Mg-1Zn-1Mn, 2Zn-1Mn, 3Zn-1Mn	uSBF	–	–	37	–	–	PDP, CTx, Hml, TT, SEM, Opt	[96]
Mg-2.7Nd-0.16Zn-0.4Zr	HBSS	–	–	37	Y, 7.4	10 days	ML, H2E, PDP, TT, Opt, XRD, SEM	[97]
Mg-2Gd, 5Gd, 10Gd, 15Gd	1 % NaCl	–	HT	22	N, 6.5	–	H2E, ML, Opt, SEM, EDS, TEM, XRD, TT, CT	[98]
Mg-2Zn-0.2Ca	–	–	±µArc, ±ED Coating	–	–	12, 18 weeks	SEM, EDS, XRD, µCT, (In Vivo)	[99]
Mg-2Zn-1Mn, 2Zn-1Mn-0.4Y, 2Zn-1Mn-0.8Y, 2Zn-1Mn-1.5Y	HBSS	–	Ext	37	N, 7.4	24 h	PDP, pHR, OCP, SEM, XRD, XPS, EDS	[100]
Mg-2Zn-1Mn-0.3Ca, 2Zn-1Mn,0.5Ca, 1.5Zn-1Mn-1Ca	HBSS	–	–	37	–	15 months	VfM, PDP, TT, SEM, EDS, ICP-AES	[101]
Mg-3Ca	0.9 % NaCl	–	Surface Roughness Modifications	20	Y (flow)	16 days	ML, H2E,	[102]
Mg-3Nd-0.27Zn-0.4Zr	uSBF	–	HT, Ext	37	Y, 7.4	10 days	ML, EIS, TT, Opt, XRD, SEM	[103]

(continued)

(continued)

Alloy	Solution	Buffer	Coating/prework	°C	pH control	Time	Experiments	References
Mg-3Nd-0.2Zn-0.4Zr	uSBF	–	HT, Ext	37	Y, 7.4	10 days	ML, H2E, EDS, CTx, Opt, XRD, SEM	[104]
Mg-4Y-2Nd	uSBF, AP	TRIS	±And	37	N	24 h	EIS, pHR, SEM, AES, Opt	[105]
Mg-6Zn	uSBF	–	–	37	N, 7.4	3, 30 d	PDP, EIS, Hml, XRD, EDS, SEM	[106]
Mg-6Zn-15Ca$_8$(PO$_4$)$_2$	–	–	–	–	–	12 weeks	Ion, μCT (In Vivo)	[107]
Mg-6Zn-xCa$_3$	uSBF	–	–	37	–	30 days	HT, SEM, EDS, CT, TT, XRD, CTx, (In Vivo)	[108]
Mg-8Y	3.5 % NaCl	–	Zone Solidified	–	–	10 months	PDP, VH, Opt, SEM, EDS, XFS	[109]
Mg-Li-Al-Re	HBSS	–	–	37	–, 7.4	3, 10 days	H2E, PDP, TT, CTx, Hml, Opt, PA, XRD, SEM	[110]
Mg-Sr (0.3–2.5 %), WE43	HBSS	–	–	37	N, 7.4	3 weeks	PDP, H2E, ML, CTx, XPS, Opt, SM, XRD, SEM, EDS (In Vivo)	[111]
Mg-Zn-Ca (–)	uSBF	–	CaP	37	–	1 h	PDP, SEM, EDS, SSRT, XRD	[112]

(continued)

(continued)

Alloy	Solution	Buffer	Coating/prework	°C	pH control	Time	Experiments	References
WE43	NaCl ($\pm$CaCl$_2$, K$_2$HPO$_4$), m-SBF ($\pm$40g/L BSA)	HPS	–	20 / 37	N, 7.4	2 weeks	PDP, EIS, OCP	[113]
WE43	NaCl ($\pm$CaCl$_2$, K$_2$HPO$_4$), m-SBF ($\pm$40g/L BSA)	HPS	Ext	37	Y, 7.4	2–5 days	SEM, Opt, EDS, XRD, FTIR	[114]
WE43	SBF with 3 Cell Lines	–	–	37	–	48 h	Cell Study	[115]
WE43	3 % NaCl, uSBF, AP	–	Ext, $\pm$ HT	20	N, 6, 7.4	24 h	EIS, TT, SEM, Opt, EDS, AES	[116]
WE43, Mg	HBSS+FBS	SB$_{CO2}$	CaHPO$_4$	37	N, 7.4	4, 8 weeks	ML, CT, SEM,	[117]
ZK60	HBSS	SB$_{CO2}$	MAO	37	Y, 7.4	5, 30 days	PDP, EIS, ML, Hml, CTx, SEM, XRD, EDS	[118]
ZW21, WZ21	uSBF	–	–	20	Y, <8	7 days	H2E, TT	[119]
ZX50	–	–	$\pm$µArc	–	–	4, 12, 24 weeks	µCT, Opt, (In Vivo)	[120]
ZX50, WZ21	–	–	–	–	–	4, 8, 12, 16, 20, 24 weeks	µCT, Opt, (In Vivo)	[121]
ZX50, WZ21, MgZnCa BMG	–	–	–	–	–	1, 3, 6 months	Opt, Pull Out, (In Vivo)	[122]

(continued)

(continued)

Alloy	Solution	Buffer	Coating/prework	°C	pH control	Time	Experiments	References
"AMS" Mg	–	–	–	–	–	1,6,13 months	Opt, (Human)	[123]
"AMS" Mg	–	–	–	–	–	6, 13 months	Opt, (H0075man)	[124]
"AMS" Mg	–	–	–	–	–	–	Opt, (Human)	[125]
"AMS" Mg	–	–	–	–	–	1, 3 months	Opt, (Human)	[126]
"AMS" Mg	–	–	–	–	–	–	Opt, (Human)	[127]
"DREAMS" Mg	–	–	–	–	–	1, 6, 24, 36 months	Opt, μCT (Human)	[128]

References

1. Feser K, Kietzmann M, Baumer W, Krause C, Bach FW (2010) Effects of degradable Mg-Ca alloys on dendritic cell function. J Biomater Appl, 0885328209360424. doi:10.1177/0885328209360424
2. Chng CB, Lau DP, Choo JQ, Chui CK (2012) A bioabsorbable microclip for laryngeal microsurgery: design and evaluation. Acta Biomater 8(7):2835–2844. doi:10.1016/j.actbio.2012.03.051
3. Pietak AM, Mahoney T, Dias G, Staiger MP (2007) Bone-like Matrix Formation on Magnesium and Magnesium Alloys. J Biomed Mater Res 19(1):407–415
4. Fischerauer SF, Kraus T, Wu X, Tangl S, Sorantin E, Hänzi AC, Löffler JF, Uggowitzer PJ, Weinberg AM (2013) In vivo degradation performance of micro-arc-oxidized magnesium implants: a micro-CT study in rats. Acta Biomater 9(2):5411–5420. doi:10.1016/j.actbio.2012.09.017
5. Vojtěch D, Čížová H, Volenec K (2006) Investigation of magnesium-based alloys for biomedical applications. Kovove Mater 44:211–223
6. Al-Abdullat Y, Tsutsumi S, Nakajima N, Ohta M, Kuwahara H, Ikeuchi K (2001) Surface modification of magnesium by NaHCO3 and corrosion behavior in Hank's solution for new biomaterial applications. Mater Trans 42(8):1777–1780
7. Yamamoto A, Hiromoto S (2009) Effect of inorganic salts, amino acids and proteins on the degradation of pure magnesium in vitro. Mater Sci Eng: C 29(5):1559–1568. doi:10.1016/j.msec.2008.12.015
8. Kuwahara H, Al-Abdullat Y, Mazaki N, Tsutsumi S, Aizawa T (2001) Precipitation of Magnesium Apatite on pure magnesium surface during immersing in hank's solution. Mater Trans 42(7):1317–1321
9. Hiromoto S, Yamamoto A, Maruyama N, Somekawa H, Mukai T (2008) Polarization behavior of pure magnesium under a controlled flow in a nacl solution. Mater Trans 49(6):1456–1461
10. Hiromoto S, Yamamoto A, Maruyama N, Somekawa H, Mukai T (2008) Influence of pH and flow on the polarisation behaviour of pure magnesium in borate buffer solutions. Corros Sci 50(12):3561–3568
11. Zhuang H, Han Y, Feng A (2008) Preparation, mechanical properties and in vitro biodegradation of porous magnesium scaffolds. Mater Sci Eng: C 28(8):1462–1466
12. Lorenz C, Brunner JG, Kollmannsberger P, Jaafar L, Fabry B, Virtanen S (2009) Effect of surface pre-treatments on biocompatibility of magnesium. Acta Biomater 5(7):2783–2789
13. Wang Y, Wei M, Gao J (2009) Improve corrosion resistance of magnesium in simulated body fluid by dicalcium phosphate dihydrate coating. Mater Sci Eng: C 29(4):1311–1316
14. Wang Y, Wei M, Gao J, Hu J, Zhang Y (2008) Corrosion process of pure magnesium in simulated body fluid. Mater Lett 62(14):2185–2188
15. Li L, Gao J, Wang Y (2004) Evaluation of cyto-toxicity and corrosion behavior of alkali-heat-treated magnesium in simulated body fluid. Surface Coat Technol 185(1).92–98
16. Kuwahara H, Al-Abdullat Y, Ohta M, Tsutsumi S, Ikeuchi K, Mazaki N, Aizawa T (2000) Surface reaction of magnesium in Hank's Solutions. In: Nagaoka City, Japan, 2000. Material Science Forum. Trans Tech Publications, pp 349–358
17. Zhang E, Xu L, Yang K (2005) Formation by ion plating of Ti-coating on pure Mg for biomedical applications. Scripta Mater 53(5):523–527
18. Lee J-Y, Han G, Kim Y-C, Byun J-Y, Jang J-i, Seok H-K, Yang S-J (2009) Effects of impurities on the biodegradation behavior of pure magnesium. Met Mater Int 15(6):955–961
19. Kannan MB (2010) Influence of microstructure on the in-vitro degradation behaviour of magnesium alloys. Mater Lett 64:739–742
20. Yun Y, Dong Z, Yang D, Schulz MJ, Shanov VN, Yarmolenko S, Xu Z, Kumta P, Sfeir C (2009) Biodegradable Mg corrosion and osteoblast cell culture studies. Materials Sci Eng: C 29(6):1814–1821

21. Inoue H, Sugahara K, Yamamoto A, Tsubakino H (2002) Corrosion rate of magnesium and its alloys in buffered chloride solutions. Corrosion Sci 44(3):603–610
22. Brar HS, Platt MO, Sarntinoranont M, Martin PI, Manuel MV (2009) Magnesium as a biodegradable and bioabsorbable material for medical implants. JOM 61(9):31–34. doi: 10.1007/s11837-009-0129-0
23. Lopez HY, Cortes DA, Escobedo S, Mantovani D (2006) In vitro bioactivity assessment of metallic magnesium. Key Eng Mater 309–311:453–456
24. Ng WF, Chiu KY, Cheng FT (2010) Effect of pH on the in vitro corrosion rate of magnesium degradable implant material. Mater Sci Eng: C 30(6):898–903
25. Chiu KY, Wong MH, Cheng FT, Man HC (2007) Characterization and corrosion studies of fluoride conversion coating on degradable Mg implants. Surface Coat Technol 202(3):590–598
26. Ng WF, Wong MH, Cheng FT (2010) Stearic acid coating on magnesium for enhancing corrosion resistance in Hanks' solution. Surface Coat Technol 204(11):1823–1830
27. Shi P, Ng WF, Wong MH, Cheng FT (2009) Improvement of corrosion resistance of pure magnesium in Hank's solution by microarc oxidation with Sol-gel TiO2 sealing. J Alloys Compd 469(1–2):286–292
28. Wan Y, Xiong G, Luo H, He F, Huang Y, Zhou X (2008) Preparation and characterization of a new biomedical magnesium-calcium alloy. Mater Des 29(10):2034–2037
29. Song G, Song S (2007) A possible biodegradable magnesium implant material. Adv Eng Maters 9(4):298–302
30. Wen Z, Wu C, Dai C, Yang F (2009) Corrosion behaviors of Mg and its alloys with different Al contents in a modified simulated body fluid. J Alloys Compd 488(1):392–399
31. Yfantis CD, Yfantis DK, Anastassopoulou J, Theophanides T, Staiger MP (2006) New magnesium alloys for bone tissue engineering: in vitro corrosion testing. WSEAS Trans Environ Develop 2(8):1110–1115
32. Mueller WD, de Mele MFL, Nascimento ML, Zeddies M (2009) Degradation of magnesium and its alloys: Dependence on the composition of the synthetic biological media. J Biomed Mater Res Part A 90A(2):487–495
33. Mueller WD, Nascimento ML, Zeddies M, Córsico M, Gassa LM, de Mele MAFL (2007) Magnesium and its alloys as degradable biomaterials: corrosion studies using potentiodynamic and EIS electrochemical techniques. Mater Res 10:5–10
34. Ren Y, Wang H, Huang J, Zhang B, Yang K (2007) Study of biodegradation of pure magnesium. Key Eng Mater 342–343:601–604
35. Walker J, Shadanbaz S, Kirkland NT, Stace E, Woodfield T, Staiger MP, Dias GJ (2012) Magnesium alloys: Predicting in vivo corrosion with in vitro immersion testing. J Biomed Mater Res Part B: Appl Biomater 100B(4):1134–1141. doi:10.1002/jbm.b.32680
36. Zhang S, Zhang X, Zhao C, Li J, Song Y, Xie C, Tao H, Zhang Y, He Y, Jiang Y, Bian Y (2010) Research of Mg-Zn alloy as degradable biomaterial. Acta Biomater 6(2):626–640
37. Kirkland NT, Lespagnol J, Birbilis N, Staiger MP (2010) A Survey of Bio-Corrosion Rates of Magnesium Alloys. Corrosion Sci 52(2):287–291. doi:10.1016/j.corsci.2009.09.033
38. Abdal-hay A, Barakat NAM, Lim JK (2013) Influence of electrospinning and dip-coating techniques on the degradation and cytocompatibility of Mg-based alloy. Colloids Surfaces A: Physicochemical Eng Aspects 420:37–45. doi:10.1016/j.colsurfa.2012.12.009
39. Abdal-hay A, Amna T, Lim JK (2013) Biocorrosion and osteoconductivity of PCL/nHAp composite porous film-based coating of magnesium alloy. Solid State Sci 18:131–140. doi: 10.1016/j.solidstatesciences.2012.11.017
40. Abdal-hay A, Dewidar M, Lim JK (2012) Biocorrosion behavior and cell viability of adhesive polymer coated magnesium based alloys for medical implants. Appl Surface Sci 261:536–546. doi:10.1016/j.apsusc.2012.08.051
41. Wang J, Zeng R, Chen J, Chen R (2009) Corrosion behaviour of magnesium alloy AZ53 in simulated body fluids. Mater Sci Forum 610–613:1174–1178

42. Yang JX, Cui FZ, Yin QS, Zhang T, Wang XM (2009) Characterization and degradation study of calcium phosphate coating on magnesium alloy bone implant in vitro. IEEE Trans Plasma Sci 37(7):1161–1168

43. Cui FZ, Yang JX, Jiao YP, Yin QS, Zhang Y (2008) Calcium phosphate coating on magnesium alloy for modification of degradation behavior. Front Mater Sci China 2(2):143–148

44. Alvarez-Lopez M, Pereda MD, del Valle JA, Fernandez-Lorenzo M, Garcia-Alonso MC, Ruano OA, Escudero ML (2010) Corrosion behaviour of AZ31 magnesium alloy with different grain sizes in simulated biological fluids. Acta Biomater 6(5):1763–1771

45. Ren Y, Huang J, Zhang B, Yang K (2007) Preliminary study of biodegradation of AZ31B magnesium alloy. Front Mater Sci China 1(4):401–404

46. Song Y, Shan D, Chen R, Zhang F, Han E-H (2009) Biodegradable behaviors of AZ31 magnesium alloy in simulated body fluid. Mater Sci Eng: C 29(3):1039–1045

47. Wang H, Estrin Y, Fu H, Song G, Zúberová Z (2007) The Effect of Pre-Processing and Grain Structure on the Bio-Corrosion and Fatigue Resistance of Magnesium Alloy AZ31. Adv Eng Mater 9(11):967–972

48. Wang H, Estrin Y, Zúberová Z (2008) Bio-corrosion of a magnesium alloy with different processing histories. Mater Lett 62(16):2476–2479

49. Gu Y, Bandopadhyay S, Chen C-f, Ning C, Guo Y (2013) Long-term corrosion inhibition mechanism of microarc oxidation coated AZ31 Mg alloys for biomedical applications. Mater Des 46:66–75. doi:10.1016/j.matdes.2012.09.056

50. 50. Krause C, Bormann D, Hassel T, Bach FW, Windhagen H, Krause A, Hackenbroich C, Meyer-Lindenberg A (2006) Mechanical properties of degradable magnesium implants in dependence of the implantation duration. In: Pekguleryuz M (ed) Conference of metallurgists: magnesium technology in the global age. Montreal, Quebec, Canada, 2006, pp 329–343

51. Zeng RC, Chen J, Dietzel W, Hort N, Kainer KU (2007) Electrochemical Behavior of Magnesium Alloys in Simulated Body Fluids. Trans Nonferrous Met Soc China 17:S166–S170

52. Mueller WD, Lucia Nascimento M, Lorenzo de Mele MF (2010) Critical discussion of the results from different corrosion studies of Mg and Mg alloys for biomaterial applications. Acta Biomater 6(5):1749–1755

53. Yan T, Tan L, Xiong D, Liu X, Zhang B, Yang K (2010) Fluoride treatment and in vitro corrosion behavior of an AZ31B magnesium alloy. Materials Science and Engineering: C 30(5):740–748

54. Kannan MB, Raman RKS (2008) In vitro degradation and mechanical integrity of calcium-containing magnesium alloys in modified-simulated body fluid. Biomaterials 29(15):2306–2314

55. Liu C, Xin Y, Tang G, Chu PK (2007) Influence of heat treatment on degradation behavior of bio-degradable die-cast AZ63 magnesium alloy in simulated body fluid. Mater Sci Eng: A 456(1–2):350–357

56. Liu C, Xin Y, Tian X, Chu PK (2007) Degradation susceptibility of surgical magnesium alloy in artificial biological fluid containing albumin. J Mater Res 22(7):1806–1814

57. Kannan MB, Raman RKS (2008) Evaluating the stress corrosion cracking susceptibility of Mg-Al-Zn alloy in modified-simulated body fluid for orthopaedic implant application. Scripta Mater 59(2):175–178

58. Fekry AM, El-Sherif RM (2009) Electrochemical corrosion behavior of magnesium and titanium alloys in simulated body fluid. Electrochim Acta 54(28):7280–7285

59. Eliezer A, Witte F (2010) Corrosion Behaviour of Magnesium Alloys in Biomedical Environments. Adv Mater Res 95:17–20

60. Wong HM, Yeung KWK, Lam KO, Tam V, Chu PK, Luk KDK, Cheung KMC (2010) A biodegradable polymer-based coating to control the performance of magnesium alloy orthopaedic implants. Biomaterials 31:2084–2096

61. Xin Y, Liu C, Huo K, Tang G, Tian X, Chu PK (2009) Corrosion behavior of ZrN/Zr coated biomedical AZ91 magnesium alloy. Surface Coat Technol 203(17–18):2554–2557
62. Xin Y, Liu C, Zhang W, Jiang J, Tang G, Tian X, Chu PK (2008) Electrochemical behavior Al2O3/Al coated surgical AZ91 magnesium alloy in simulated body fluids. J Electrochem Soc 155(5):178–182
63. Xin Y, Liu C, Zhang X, Tang G, Tian X, Chu PK (2007) Corrosion behavior of biomedical AZ91 magnesium alloy in simulated body fluids. J Mater Res 22(7):2004–2011
64. Witte F, Feyerabend F, Maier P, Fischer J, Stormer M, Blawert C, Dietzel W, Hort N (2007) Biodegradable magnesium-hydroxyapatite metal matrix composites. Biomaterials 28(13):2163–2174
65. Levesque J, Hermawan H, Dube D, Mantovani D (2008) Design of a pseudo-physiological test bench specific to the development of biodegradable metallic biomaterials. Acta Biomater 4(2):284–295
66. Song YW, Shan DY, Han EH (2008) Electrodeposition of hydroxyapatite coating on AZ91D magnesium alloy for biomaterial application. Mater Lett 62(17–18):3276–3279
67. Zhou W, Shen T, Aung NN (2010) Effect of heat treatment on corrosion behaviour of magnesium alloy AZ91D in simulated body fluid. Corrosion Sci 52(3):1035–1041
68. Wang YM, Guo JW, Shao ZK, Zhuang JP, Jin MS, Wu CJ, Wei DQ, Zhou YA (2012) Metasilicate-based ceramic coating formed on magnesium alloy by microarc oxidation and its corrosion in simulated body fluid. Surface Coat Technol 219:8–14. doi:10.1016/j.surfcoat.2012.12.040
69. Choudhary L, Singh Raman RK (2012) Magnesium alloys as body implants: fracture mechanism under dynamic and static loadings in a physiological environment. Acta Biomater 8(2):916–923. doi:10.1016/j.actbio.2011.10.031
70. Witte F, Nellesen J, Crostack H-A, Kaese V, Pisch A, Beckmann F, Windhagen H (2006) In vitro and in vivo corrosion measurements of magnesium alloys. Biomaterials 27(7):1013–1018. doi:10.1016/j.biomaterials.2005.07.037
71. Cao JD, Kirkland NT, Laws KJ, Birbilis N, Ferry M (2012) Ca–Mg–Zn bulk metallic glasses as bioresorbable metals. Acta Biomater 8(6):2375–2383. doi:10.1016/j.actbio.2012.03.009
72. Song G (2007) Control of biodegradation of biocompatible magnesium alloys. Corrosion Sci 49(4):1696–1701
73. Wang H, Shi ZM, Yang K (2008) Magnesium and Magnesium Alloys as Degradable Metallic Biomaterials. Adv Mater Res 32:207–210
74. Gu X, Zheng Y, Zhong S, Xi T, Wang J, Wang W (2010) Corrosion of, and cellular responses to Mg-Zn-Ca bulk metallic glasses. Biomaterials 31(6):1093–1103. doi:10.1016/j.biomaterials.2009.11.015
75. Huang L, Qiao D, Green BA, Liaw PK, Wang J, Pang S, Zhang T (2009) Bio-corrosion study on zirconium-based bulk-metallic glasses. Intermetallics 17(4):195–199
76. Zberg B, Uggowitzer PJ, Loffler JF (2009) MgZnCa glasses without clinically observable hydrogen evolution for biodegradable implants. Nature Mater 8(11):887–891
77. Li Y, Wen C, Mushahary D, Sravanthi R, Harishankar N, Pande G, Hodgson P (2012) Mg–Zr–Sr alloys as biodegradable implant materials. Acta Biomater 8(8):3177–3188. doi:10.1016/j.actbio.2012.04.028
78. Hassel T, Bach FW, Krause C (2007) Influence of the alloy composition on the mechanical and electrochemical properties of binary Mg-Ca alloys and its corrosion behaviour in solutions at different chloride concentrations. In: Kainer KU (ed) Magnesium: proceedings of the 7th international conference on magnesium alloys and their applications, 2007. Wiley-VCH, pp 789–795

79. Hassel T, Bach FW, Golovko AN, Krause A (2006) Investigation of the mechanical properties and the corrosion behaviour of low alloyed magnesium-calcium-alloys for use as absorbable biomaterial in the implant technique. In: Pekguleryuz M (ed) Conference of Metallurgists : magnesium technology in the global age. Montreal, Quebec, Canada, 2006. pp 359–369

80. Denkena B, Podolsky C, Lucas A, Hassel T, Witte F, Palm O, Hurschler C (2005) Degradable implants made of magnesium alloys, vol 3. In: 5th Euspen international conference. Montpellier, France, May 2005.

81. Wan YZ, Xiong GY, Luo HL, He F, Huang Y, Wang YL (2008) Influence of zinc ion implantation on surface nanomechanical performance and corrosion resistance of biomedical magnesium-calcium alloys. Appl Surface Sci 254(17):5514–5516

82. Erdmann N, Angrisani N, Reifenrath J, Lucas A, Thorey F, Bormann D, Meyer-Lindenberg A (2011) Biomechanical testing and degradation analysis of MgCa0.8 alloy screws: a comparative in vivo study in rabbits. Acta Biomater 7(3):1421–1428. doi: 10.1016/j.actbio.2010.10.031

83. Kim W-C, Kim J-G, Lee J-Y, Seok H-K (2008) Influence of Ca on the corrosion properties of magnesium for biomaterials. Materials Lett 62(25):4146–4148

84. Xu L, Pan F, Yu G, Yang L, Zhang E, Yang K (2009) In vitro and in vivo evaluation of the surface bioactivity of a calcium phosphate coated magnesium alloy. Biomaterials 30(8):1512–1523

85. Xin Y, Huo K, Tao H, Tang G, Chu PK (2008) Influence of aggressive ions on the degradation behavior of biomedical magnesium alloy in physiological environment. Acta Biomater 4(6):2008–2015

86. Xu L, Zhang E, Yang K (2009) Phosphating treatment and corrosion properties of Mg–Mn–Zn alloy for biomedical application. J Mater Sci: Mater Med 20(4):859–867

87. Pérez P, Onofre E, Cabeza S, Llorente I, del Valle JA, García-Alonso MC, Adeva P, Escudero ML (2013) Corrosion behaviour of Mg–Zn–Y–Mischmetal alloys in phosphate buffer saline solution. Corrosion Sci 69:226–235. doi:10.1016/j.corsci.2012.12.007

88. Yang L, Hort N, Laipple D, Höche D, Huang Y, Kainer KU, Willumeit R, Feyerabend F (2012) Element distribution in the corrosion layer and cytotoxicity of alloy Mg–10Dy during in vitro biodegradation. Acta Biomater. doi:10.1016/j.actbio.2012.10.001

89. Yfantis CD, Yfantis DK, Anastassopoulou J, Theophanides T, Staiger MP (2006) In vitro corrosion behaviour of new magnesium alloys for bone regeneration.

90. Li Z, Gu X, Lou S, Zheng Y (2008) The development of binary Mg-Ca alloys for use as biodegradable materials within bone. Biomaterials 29(10):1329–1344

91. Zheng YF, Gu XN, Xi YL, Chai DL (2010) In vitro degradation and cytotoxicity of Mg/Ca composites produced by powder metallurgy. Acta Biomater 6:1783–1791

92. Zhang CY, Zeng RC, Liu CL, Gao JC (2010) Comparison of calcium phosphate coatings on Mg-Al and Mg-Ca alloys and their corrosion behavior in Hank's solution. Surface Coat Technol 204(21–22):3636–3640

93. Gu XN, Zheng YF, Chen LJ (2009) Influence of artificial biological fluid composition on the biocorrosion of potential orthopedic Mg-Ca, AZ31, AZ91 alloys. Biomed Mater 4(6):8. doi:065011 10.1088/1748-6041/4/6/065011

94. Yang L, Zhang E (2009) Biocorrosion behavior of magnesium alloy in different simulated fluids for biomedical application. Materials Sci Eng: C 29(5):1691–1696

95. Gu X, Zheng Y, Cheng Y, Zhong S, Xi T (2009) In vitro corrosion and biocompatibility of binary magnesium alloys. Biomaterials 30(4):484–498

96. Wang J, He Y, Maitz MF, Collins B, Xiong K, Guo L, Yun Y, Wan G, Huang N (2013) A surface-eroding poly(1,3-trimethylene carbonate) coating for fully-biodegradable magnesium-based stent applications: toward better biofunction, biodegradation, and biocompatibility. Acta Biomater. doi:10.1016/j.actbio.2013.02.041

97. Zhang E, Yin D, Xu L, Yang L, Yang K (2009) Microstructure, mechanical and corrosion properties and biocompatibility of Mg-Zn-Mn alloys for biomedical application. Mater Sci Eng C 29(3):987–993

98. Zhang X, Yuan G, Wang Z (2012) Mechanical properties and biocorrosion resistance of Mg-Nd-Zn-Zr alloy improved by cyclic extrusion and compression. Mater Lett 74:128–131. doi:10.1016/j.matlet.2012.01.086

99. Hort N, Huang Y, Fechner D, Störmer M, Blawert C, Witte F, Vogt C, Drücker H, Willumeit R, Kainer KU, Feyerabend F (2010) Magnesium alloys as implant materials—principles of property design for Mg-RE alloys. Acta Biomater 6:1714–1725

100. Chen S, Guan S, Li W, Wang H, Chen J, Wang Y, Wang H (2012) In vivo degradation and bone response of a composite coating on Mg–Zn–Ca alloy prepared by microarc oxidation and electrochemical deposition. J Biomed Mater Res Part B: Appl Biomater 100B(2):533–543. doi:10.1002/jbm.b.31982

101. He W, Zhang E, Yang K (2009) Effect of Y on the bio-corrosion behavior of extruded Mg-Zn-Mn alloy in Hank's solution. Materi Sci Eng: C 30:167–174

102. Zhang E, Yang L (2008) Microstructure, mechanical properties and bio-corrosion properties of Mg-Zn-Mn-Ca alloy for biomedical application. Mater Sci Eng: A 497(1–2):111–118

103. Denkena B, Lucas A (2007) Biocompatible magnesium alloys as absorbable implant materials—adjusted surface and subsurface properties by machining processes. CIRP Annals-Manuf Technol 56(1):113–116

104. Zhang X, Yuan G, Mao L, Niu J, Fu P, Ding W (2012) Effects of extrusion and heat treatment on the mechanical properties and biocorrosion behaviors of a Mg–Nd–Zn–Zr alloy. J Mech Beh Biomed Mater 7:77–86. doi:10.1016/j.jmbbm.2011.05.026

105. Zhang X, Yuan G, Niu J, Fu P, Ding W (2012) Microstructure, mechanical properties, biocorrosion behavior, and cytotoxicity of as-extruded Mg–Nd–Zn–Zr alloy with different extrusion ratios. J Mech Behav Biomed Mater 9:153–162. doi:10.1016/j.jmbbm.2012.02.002

106. Quach N-C, Uggowitzer PJ, Schmutz P (2008) Corrosion behaviour of an Mg-Y-RE alloy used in biomedical applications studied by electrochemical techniques. Comptes Rendus Chimie 11(9):1043–1054

107. Zhang S, Li J, Song Y, Zhao C, Zhang X, Xie C, Zhang Y, Tao H, He Y, Jiang Y, Bian Y (2009) In vitro degradation, hemolysis and MC3T3-E1 cell adhesion of biodegradable Mg-Zn alloy. Mater Sci Eng: C 29(6):1907–1912

108. Yu K, Chen L, Zhao J, Wang R, Dai Y, Huang Q (2013) In vivo biocompatibility and biodegradation of a Mg-15%Ca3(PO4)2 composite as an implant material. Mater Lett 98:22–25. doi:10.1016/j.matlet.2013.02.018

109. Yu K, Chen L, Zhao J, Li S, Dai Y, Huang Q, Yu Z (2012) In vitro corrosion behavior and in vivo biodegradation of biomedical β-Ca3(PO4)2/Mg–Zn composites. Acta Biomater 8(7):2845–2855. doi:10.1016/j.actbio.2012.04.009

110. Peng Q, Huang Y, Zhou L, Hort N, Kainer KU (2010) Preparation and properties of high purity Mg-Y biomaterials. Biomaterials 31(3):398–403

111. Zhou WR, Zheng YF, Leeflang MA, Zhou J (2013) Mechanical property, biocorrosion and in vitro biocompatibility evaluations of Mg–Li–(Al)–(RE) alloys for future cardiovascular stent application. Acta Biomater. doi:10.1016/j.actbio.2013.01.032

112. Bornapour M, Muja N, Shum-Tim D, Cerruti M, Pekguleryuz M (2013) Biocompatibility and biodegradability of Mg–Sr alloys: The formation of Sr-substituted hydroxyapatite. Acta Biomater 9(2):5319–5330. doi:10.1016/j.actbio.2012.07.045

113. Wang HX, Guan SK, Wang X, Ren CX, Wang LG (2010) In vitro degradation and mechanical integrity of Mg-Zn-Ca alloy coated with Ca-deficient hydroxyapatite by the pulse electrodeposition process. Acta Biomater 6:1743–1748

114. Rettig R, Virtanen S (2008) Time-dependent electrochemical characterization of the corrosion of a magnesium rare-earth alloy in simulated body fluids. J Biomed Mater Res Part A 85A(1):167–175

115. Rettig R, Virtanen S (2009) Composition of corrosion layers on a magnesium rare-earth alloy in simulated body fluids. J Biomed Mater Res—Part A 88(2):359–369

116. Feyerabend F, Fischer J, Holtz J, Witte F, Willumeit R, Drücker H, Vogt C, Hort N (2010) Evaluation of short-term effects of rare earth and other elements used in magnesium alloys on primary cells and cell lines. Acta Biomater 6(5):1834–1842

117. Gunde P, Angela F, Anja CH, Patrik S, Peter JU (2010) The influence of heat treatment and plastic deformation on the bio-degradation of a Mg-Y-RE alloy. J Biomed Mater Res Part A 92A(2):409–418

118. Liao Y, Chen D, Niu J, Zhang J, Wang Y, Zhu Z, Yuan G, He Y, Jiang Y (2012) In vitro degradation and mechanical properties of polyporous CaHPO4-coated Mg–Nd–Zn–Zr alloy as potential tissue engineering scaffold. Mater Lett 100:306–308. doi:10.1016/j.matlet.2012.09.119

119. Lin X, Tan L, Zhang Q, Yang K, Hu Z, Qiu J, Cai Y (2012) The in vitro degradation process and biocompatibility of a ZK60 magnesium alloy with a forsterite-containing micro-arc oxidation coating. Acta Biomater (0). doi:10.1016/j.actbio.2012.12.016

120. Hänzi AC, Weder MM, Gerold B, Uggowitzer PJ (2007) New bio-absorbable magnesium alloys for medical applications. In: Light Metals technology conference, Canada, 2007

121. Kraus T, Fischerauer SF, Hänzi AC, Uggowitzer PJ, Löffler JF, Weinberg AM (2012) Magnesium alloys for temporary implants in osteosynthesis: in vivo studies of their degradation and interaction with bone. Acta Biomater 8(3):1230–1238. doi:10.1016/j.actbio.2011.11.008

122. Celarek A, Kraus T, Tschegg EK, Fischerauer SF, Stanzl-Tschegg S, Uggowitzer PJ, Weinberg AM (2012) PHB, crystalline and amorphous magnesium alloys: promising candidates for bioresorbable osteosynthesis implants? Mater Sci Eng: C 32(6):1503–1510. doi:10.1016/j.msec.2012.04.032

123. Bosiers M (2009) AMS INSIGHT—Absorbable metal stent implantation for treatment of below-the-knee critical limb ischemia: 6-month analysis. Cardiovasc Intervent Radiol 32(3):424–435. doi:10.1007/s00270-008-9472-8

124. Erbel R, Di Mario C, Bartunek J, Bonnier J, de Bruyne B, Eberli FR, Erne P, Haude M, Heublein B, Horrigan M, Ilsley C, Böse D, Koolen J, Lüscher TF, Weissman N, Waksman R (2007) Temporary scaffolding of coronary arteries with bioabsorbable magnesium stents: a prospective, non-randomised multicentre trial. Lancet 369(9576):1869–1875. doi:http://dx.doi.org/10.1016/S0140-6736(07)60853-8

125. McMahon CJ, Oslizlok P, Walsh KP (2007) Early restenosis following biodegradable stent implantation in an aortopulmonary collateral of a patient with pulmonary atresia and hypoplastic pulmonary arteries. Cathet Cardiovasc Interv 69(5):735–738. doi:10.1002/ccd.21091

126. Peeters P, Bosiers M, Verbist J, Deloose K, Heublein B (2005) Preliminary results after application of absorbable metal stents in patients with critical limb ischemia. J Endovasc Ther 12(1):1–5. doi:10.1583/04-1349R.1

127. Schranz D, Zartner P, Michel-Behnke I, Akintürk H (2006) Bioabsorbable metal stents for percutaneous treatment of critical recoarctation of the aorta in a newborn. Cathet Cardiovasc Interv 67(5):671–673. doi:10.1002/ccd.20756

128. Haude M, Erbel R, Erne P, Verheye S, Degen H, Böse D, Vermeersch P, Wijnbergen I, Weissman N, Prati F, Waksman R, Koolen J (2013) Safety and performance of the drug-eluting absorbable metal scaffold (DREAMS) in patients with de-novo coronary lesions: 12 month results of the prospective, multicentre, first-in-man BIOSOLVE-I trial. The Lancet 381(9869):836–844. doi:10.1016/S0140-6736(12)61765-6

Index

N. T. Kirkland and N. Birbilis, *Magnesium Biomaterials*, SpringerBriefs in Materials, 131
DOI: 10.1007/978-3-319-02123-2, © The Author(s) 2014